WILD
ANIMALS

OF NORTH AMERICA

Prepared by
The Book Division
National Geographic Society
Washington, D.C.

Snarling jaguar; *previous page:* Hibernating Indiana bats

WILD ANIMALS
OF NORTH AMERICA

Published by
The National Geographic Society

John M. Fahey, Jr.
President and Chief Executive Officer
Gilbert M. Grosvenor
Chairman of the Board
Nina D. Hoffman
Senior Vice President

Prepared by The Book Division
William R. Gray
Vice President and Director
Charles Kogod
Assistant Director
Barbara A. Payne
Editorial Director and Managing Editor

Staff for This Book
Tom Melham
Managing Editor
Marilyn Gibbons
Illustrations Editor
Lyle Rosbotham
Art Director
Elisabeth B. Booz,
Elizabeth Cook Thompson
Researchers
R. Gary Colbert
Production Director
Lewis R. Bassford
Production Project Manager
Richard S. Wain
Production
Meredith C. Wilcox
Illustrations Assistant
Peggy Candore, Dale-Marie Herring
Staff Assistants

Anne Marie Houppert
Indexer

Manufacturing and Quality Control
George V. White
Director
John T. Dunn
Associate Director
Vincent P. Ryan, Gregory Storer
Managers
Polly P. Tompkins
Executive Assistant

 Library of Congress CIP Data: page 200

Beluga whales approach for a closer look

Contents

Bull elk in velvet takes in a Wyoming summer

Mammals

Class Mammalia

BY RONALD M. NOWAK

Headlines of October 1996 proclaimed the bad news: a quarter of all species of mammals faced extinction. The prestigious Swiss-based International Union for the Conservation of Nature (IUCN) had just released its latest *Red List of Threatened Animals*. U.S. Interior Secretary Bruce Babbit quickly termed the inch-thick report, which drew on a global network of over 6,000 experts, "probably the most thorough scientific assessment of the state of the world's wildlife ever." Careful review of the document would show that the alarming headlines actually understated the crisis. When subspecies—that is, geographic races—and "near-threatened" mammals were added in, the number in jeopardy was closer to half the world's total of 4,600 species.

How could this be? Mammals are highly evolved vertebrates (animals with backbones); they include the most intelligent, adaptable, and socially organized creatures on land or sea, as well as the biggest. Yet a greater proportion of them are in trouble than are non-mammals. Unfortunately, size and complexity may increase vulnerability; the same factors that have made mammals so successful could lead to their demise. To unravel this paradox, we must go back to the origins of this group.

The word "mammal" comes from the Latin *mamma*, referring to the breast or nipple through which the female passes milk to her young. The very sound of that Latin term has been incorporated in languages worldwide as the term for a female parent. All mammals, except for the primitive monotremes—the platypus and spiny anteaters of Australia and New Guinea—have at least one pair of nipples. Even monotremes have mammary glands; their young lap milk as it seeps through pores in the abdominal skin. They also lay eggs, as did the earliest mammals, which evolved from reptiles around 200 million years ago. Australian biologist Marilyn Renfree suggests that lactation may have begun as an antibacterial secretion to protect eggs and hatchlings, gradually evolving into a flow of nutrients. The nourishment of offspring through mother's milk unites all mammals and sets them apart from other animal groups. Direct dependence of the young on the body of the mother, even after they are born, tends to lengthen affinity between individuals, and hence to develop social relations, communication, and the transmission of knowledge from one generation to another. Yet, while this process may assure fitter individuals at maturity, it also draws out the fragile period when young are susceptible to direct harm and to loss of the mother. A process millions of years in the making can be shattered by the sudden intrusion of unnatural factors.

Second to lactation, the most apparent characteristic of mammals is hair, a nonliving derivative of the outermost layer of skin. Hair's primary function is to insulate by trapping air and retarding loss of body heat. No mammal goes an entire lifetime without some hair, though in whales and porpoises it may be present only in the embryonic state or as just a few bristles around the mouths of adults.

An insulating coat is one of a suite of features enabling mammals to follow a more vigorous, adaptive life than that of their reptilian ancestors. Another mammalian feature concerns the heart, which is fully separated into left and right sections. When freshly aerated blood from the lungs enters the left side of the heart to be pumped throughout the body, it cannot mix with deoxygenated blood returning to the right side for pumping back to the lungs—as it can in reptiles and amphibians. This means a more efficient supply of oxygen to the tissues, which makes possible higher metabolic rates and more available energy. The resultant heat, retained through natural insulation, permits the mammalian body to keep its temperature relatively constant, regardless of outside conditions. Mammals thus can survive frigid climates, can function on cold nights, can run for long distances, and can incubate their eggs internally until the young are born.

Mammalian teeth, unlike those of other animals, are differentiated, usually into four types: incisors, canines, premolars, and molars. This empowers mammals to seize, cut, slice, crush, and grind their food, delivering it to the stomach in a form that can be more easily digested and used to maintain high energy levels. Such advanced dentition evolved with a stronger, more efficient arrangement of the jaws. In other vertebrates, each side of the lower jaw relies on several different bones. Mammals possess a single jawbone, hinged directly to the skull.

During the transition from reptile to mammal, creatures' limbs came to extend downward from the body rather than outward, thus facilitating more rapid movement. The fastest mammals, such as wild dogs and cats, have feet with heels raised well above the ground. Horses, deer, and many other large plant eaters run on the tips of their toes, on large protective nails we call hoofs. The kangaroos of Australia and the unrelated kangaroo rats of North America have relatively huge hind feet that allow a rapid, bounding gait. Moles and gophers, which spend much of their lives tunneling through soil, have big and powerful front feet and claws.

Many mammals can climb, aided by flexible digits and well developed claws. Squirrels are our most arboreal mammals. Flying squirrels don't actually fly, but they can glide by spreading open loose folds of skin that extend from wrists to ankles. The only mammals capable of true (flapping) flight are bats, which use this skill to chase insects and reach otherwise inaccessible food and roosts. Some bat species depend upon forests both for roosts and food, while others roost in forests and fly to surrounding areas to feed. Still others roost in caves and commute to foraging areas in forests or fields. Bat species that rely on caves may be more vulnerable to disturbance by people.

Some mammals, such as beavers and muskrats, have adapted to a semiaquatic life by developing webbed feet and flattened tails. Seals and sea lions have finlike feet. The tail of the manatee has been modified into a large paddle, while its hind limbs have disappeared. Whales, dolphins, and porpoises have become so adapted to moving through water that they can be mistaken for fish. All marine mammals, however, must surface to breathe; seals also must venture onto land or ice floes to give birth. Therefore, despite their size, strength, intelligence, and remote and vast habitats, these commercially valuable, water-dwelling creatures are among the mammals most susceptible to the impacts of humans.

Class Mammalia Mammals

Various land mammals also have been hunted extensively, but now many are jeopardized mainly by habitat destruction. Some, especially smaller species, are adapted to areas with particular vegetation, moisture levels, soil types, and temperature ranges. In California and the Southwest, for example, suitable water and cover are spotty at best—and are being increasingly fragmented by human activity. Here we find the nation's greatest concentration of threatened wildlife, notably many varieties of mice, voles, chipmunks, kangaroo rats, and shrews.

The area normally used by an animal is its home range—less than half an acre for a mouse or as large as 5,000 square miles for a wolf pack. It frequently contains particular places for shelter, rest, or giving birth. Among rodents, for example, we find the elaborate burrows of prairie dogs, the suspended nests of golden mice, and the complex dens of beavers. A home range may include a territory, a zone that an individual or social group defends against others of its own kind.

Mountain lions mark large territories, seemingly satisfied never to see one another except for mating and maternal care. The woodchuck also is solitary. But its western cousin, the Olympic marmot, is a sociable rodent that forms large colonies based on a mated pair and successive litters of offspring. Strong bonding between several generations of females is a basis of group formation in mammals as diverse as ground squirrels, coatis, mountain sheep, and sperm whales. Young males, usually less amicable than females, sometimes form their own groups. These aggregations, such as those of fur seals and pronghorn, may quickly break down as maturity approaches and the animals begin to compete for mating privileges.

Whatever social system a species has evolved, there must be adequate space and resources to allow for territoriality, mating rituals, and rearing of young. As more areas are disturbed, altered, reduced, and even eliminated by human activity, more and more mammals will join the list of the threatened. ■

Editor's note: Taxonomists, those biologists who specialize in categorizing life forms, seek not merely to subdivide, but also to decipher and define the myriad branches, branchlets, and twigs of life's endlessly varied family tree. To do so, they separate the animal kingdom into distinct groups, called phyla, which in turn are subdivided into classes. Each class is split into separate orders, each order into different families, each family into various genera. A single genus may include dozens, even hundreds, of different species—or only one.

This book, which concentrates on the class called mammals, follows current taxonomic lines; each chapter focuses on a single order, briefly discussing its North American families and selected species. A complete list of the continent's living mammals can be found on pages 192-195. Taxonomy, however, is not a static pursuit. Its categories are in continual flux, changing as our perceptions of our fellow creatures change. New studies of anatomical structures, cellular organization, or even DNA analysis can alter accepted taxonomic lines. This book's list follows today's rules; tomorrow's discoveries inevitably will require some adjustment.

Sleeping gray squirrel displays one mammalian trademark: hair

Marsupials

BY DON E. WILSON

Say the word "marsupial" and people immediately think of kangaroos and koala bears. Actually, this group is so diverse that biologists who specialize in the study of classification have recently elevated Marsupalia from an order to a subclass of the class Mammalia. We now recognize seven separate orders within that subclass, four of which are restricted to Australia, two to South America. Only one, the order Didelphimorphia, extends from South America into North America. And only one species—the familiar Virginia opossum, *Didelphis virginiana*—has successfully penetrated north of the U.S.-Mexico border.

This relatively recent arrival has a long and distinguished fossil ancestry, with forebears extending back into the Cretaceous period, some 100 million years ago. Interestingly enough, marsupials probably originated in North America, radiating to Australia via South America and Antarctica when those continents were conjoined. Early marsupials seem to have died out in North America by about 20 million years ago. But during the Pliocene epoch, which began some 5 million years ago, the appearance of a land bridge linking the Americas allowed some marsupial groups to invade our continent from the south.

Today, several species of the family Didelphidae occur in Mexico and Central America, but the Virginia opossum is the only one to adapt to North America's colder climates. These creatures, while slow moving, are actually remarkably well adapted to change. Unlike most other mammals, they find the company of humans quite compatible with their scavenging lifestyle. When Europeans first arrived in North America, opossums were not known to occur north of Pennsylvania. They have since expanded their range, appearing as far north as Ontario by 1858, and following agricultural expansion westward through the Great Plains. Introduced into California, they eventually spread all along the West Coast.

"Marsupial" stems from "marsupium," the abdominal pouch that characterizes females of the group. The pouch contains the animal's mammae, which vary considerably in number and placement among different species. Virginia opossums have 13 mammae, arranged in a circular pattern with a single one in the middle. Female opossums go into estrus every 28 days, much as humans do. But they may have two or even three litters per year, in some areas. The gestation period is a remarkably short 12 to 13 days, and the exceptionally immature young scramble into the pouch and attach themselves to a nipple. They are tiny indeed, less than half an inch long and weighing about 200 to the ounce! Litters can be huge, with 20 or more not uncommon and one report of 56 for a single litter. In order to survive, however, the young must attach to a nipple almost immediately after birth.

Those that make it into the pouch and find a nipple remain attached for 50 to 60 days, while development proceeds much as it does internally in placental mammals. The young first venture outside the pouch at around 70 days and are completely weaned by 100 days. As the rapidly growing youngsters overflow the pouch, they clamber about the mother's fur, even riding on her back from time to time. They wander off shortly after weaning, and reach sexual maturity at about

previous pages: **Virginia opossum clings to branches**

six to eight months. Females breed for only a couple of years, rarely living beyond three years in the wild, although they may survive for up to five years in captivity.

Opossums occur in a wide variety of habitats but usually prefer woods. They are active primarily at night, retreating by day to dens located in hollow trees, in rocky crevices, or even in underground burrows. They furnish their dwellings with leaves, twigs, and grass, which they transport either by mouth or by using their prehensile tail. Common among some arboreal mammals, prehensile tails further attest to the versatility of this animal, which is equally at home on the ground or in bushes and trees. Opossums often establish regular pathways between their daytime retreats and favorite foraging areas. One summer my family and I watched a hefty male climb the lilac bush outside our suburban Virginia home to gain access to the roof, which allowed him to continue on through some large trees in the back yard. He used this route several times a week for a couple of months.

Although their slow, shambling gait makes opossums seem a bit clumsy on the ground, they are strong climbers and swimmers. Also inveterate scavengers, they feed on virtually anything they can catch or scrounge, including small vertebrates, insects, carrion, and many kinds of fruits and vegetable matter. They do not hibernate, even in the far north, but they do store large amounts of body fat, and may stay in their dens for several days at a time to wait out inclement weather.

Opossums normally have relatively low population densities, averaging about one animal for every ten acres. Each may occupy a home range ten times as large, foraging for over a mile each night. They are somewhat nomadic, occupying a favored site for only a few months before moving on. Although not territorial, they will actively defend their immediate space and often react aggressively to other individuals. This aggression extends across species lines, as I discovered one night at the Smithsonian Tropical Research Institute on Barro Colorado Island, in Panama. I happened upon a small *Didelphis marsupialis* confronting another species, *Philander opossum*, the gray four-eyed opossum. This particular encounter ended with the larger *Didelphis* making a meal of the *Philander*.

Virginia opossums are known for a familiar behavioral trait called "playing 'possum." It is actually a physiological state of involuntary catatonia, a reflex reaction that, like human fainting, results in the animal becoming immobile, usually with the mouth open, seemingly oblivious to all outside stimuli. It seems to be a predator-avoidance strategy, based on the idea that lack of motion can cause a potential predator to lose interest. Such traits are favored by natural selection even if they are only slightly more successful than chance alone. In a slow-moving and rather defenseless animal like the opossum, becoming completely immobile may well be more effective than trying to outrun a much faster predator.

Although we often think of marsupials as "primitive" animals, the remarkable versatility of the Virginia opossum has served it well for thousands of years. By being a generalist rather than specializing to fit some narrow niche, this species has increased its capacity to adapt to the rapidly changing world in which it lives. ■

Youngsters encounter a no-passing zone

Family Didelphidae   New World Opossums

Virginia Opossum (*Didelphis virginiana*) Early speculation about these curious animals that carry their young in an abdominal pouch resulted in some strange myths, a few of which continue to crop up from time to time. The animal's forked penis may have led to conjecture that males copulate through the female's nostrils and that the young are subsequently blown into the pouch. Actually, opossum reproduction mirrors that of placental mammals, with the exception that the young are born after an extremely short gestation of less than two weeks and spend most of their early developmental period in the pouch.

Young opossums, like all mammals, rely on their mother for milk until they are weaned and able to forage on their own. They remain with her for their first hundred or so days of life, seldom venturing outside the pouch until at least two months old. For the next few months, they continue to stick close to mom, even when in the den, which may be comfortably furnished with leaves, grass, or other soft material. Even immatures are adept at climbing, using their prehensile tails as safety lines on narrow branches. Equally important are the opposable first toes on their hind feet, which allow them to grasp narrow branches and scramble through low bushes, as well as climb more substantial trees. Adult males have been observed marking objects with their saliva, suggesting a territorial behavior or at least advertisement of their home range. Males react aggressively when they encounter each other unexpectedly, often hissing and screeching. Yet these contemporary descendants of an ancient lineage are adaptable mammals whose ability to coexist with humans argues for a promising future.

Home sweet home

Insectivores

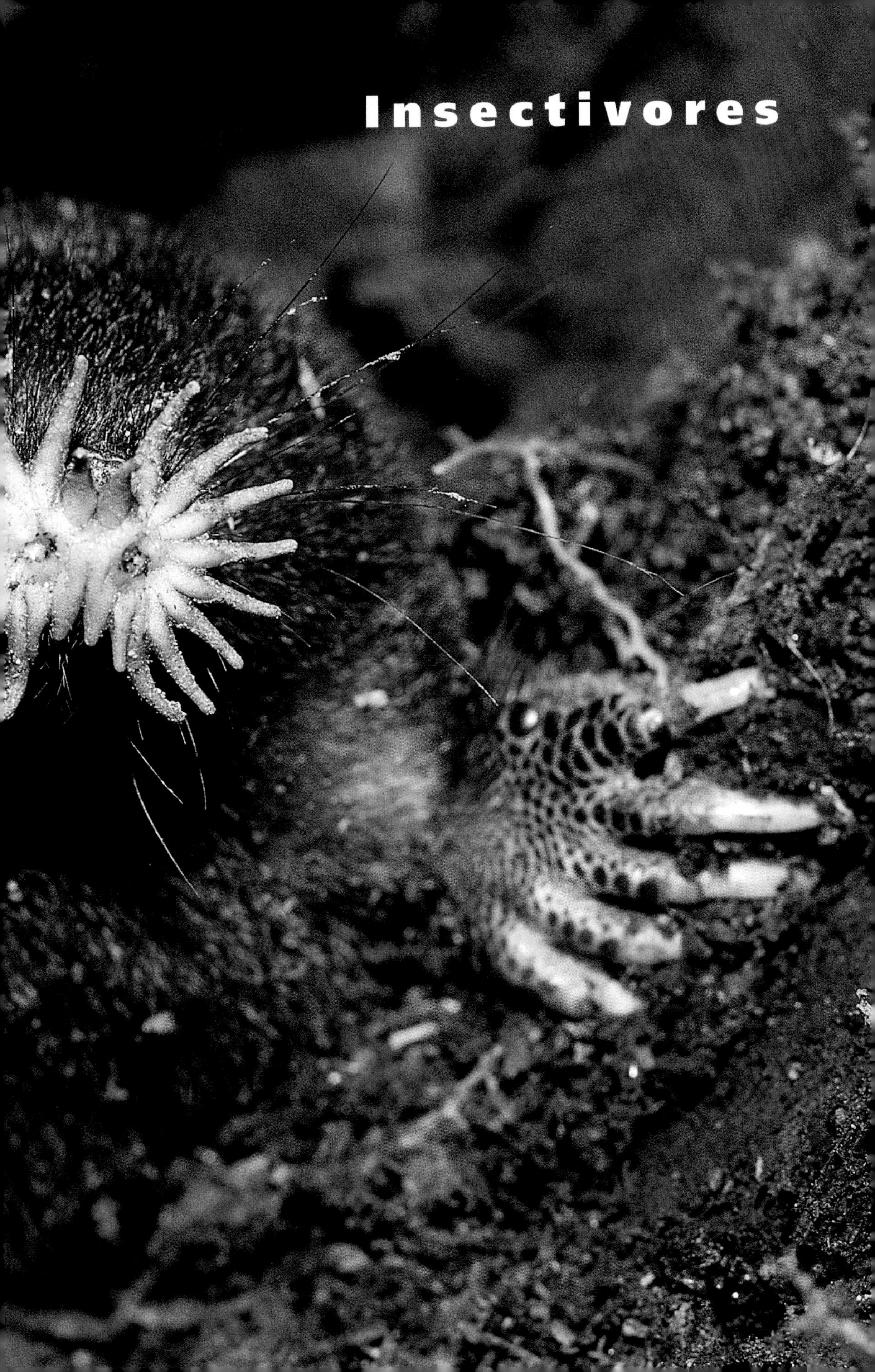

Order Insectivora

BY DON E. WILSON

Insectivores—shrews, moles, and their relatives—share many anatomical similarities, as well as a general tendency to feed on small invertebrates, especially insects. Worldwide, this group ranks as the third largest mammalian order, embracing 440 living species, 68 genera, and 6 families. North America harbors 40 species: 7 moles (family Talpidae) and 33 shrews (family Soricidae). Widely distributed throughout the continent, insectivores range from southwestern deserts to Canada's Maritime Provinces, from the Rocky Mountains to Alaska's far north.

Shrews are relatively innocuous little creatures, rarely coming into contact with humans. Yet the term "shrew," from the Old English *screawa*, meaning animal, has taken on some thoroughly negative connotations over the years. "Shrewish" and "shrewd," for example, both convey a somewhat devious or nefarious character. Granted, shrews *are* pint-size bundles of energy, forced by small size and the resultant high metabolic rates to forage enthusiastically for a wide variety of prey. Ounce for ounce, they are among the most ferocious of animals.

Moles, on the other hand, have had a much more benign reputation, at least in children's literature and folklore. In reality, they are somewhat of a bane for gardeners and greenskeepers, due to their habit of excavating tunnels. Such activities can cause considerable soil disruption, and in some agricultural areas moles are viewed as pests. Even so, their tunneling habits are beneficial in that they help break up and aerate compacted soil. Moles also feed on a variety of insect pests, such as Japanese beetle larvae and other underground grubs.

Fossil insectivores occurred back into the Cretaceous period, and their living relatives certainly resemble the ancient forebears in their basic body plan. Popular scenarios of the dawn of the age of mammals often depict great lumbering dinosaurs being replaced by beady-eyed little shrewlike mammals. Living shrews, however, are not very closely related to those ancestral forms. Modern moles and shrews probably originated in Europe, spreading from there to North America and Asia. Europe today has two species of desmans, close relatives of moles, which have become largely aquatic. Other living families of insectivores include the hedgehogs of Eurasia and Africa; the tenrecs, water shrews, and golden moles of Africa; and the solenodons of the West Indies.

Frequently confused with rodents, shrews differ in their long, slender noses and mouthfuls of sharp little teeth—which do not grow continuously, as the incisors of rodents do. They are the smallest of terrestrial mammals, some species weighing less than a tenth of an ounce. Such small size leaves them susceptible to extremes of temperature and dehydration. Still, desert-dwelling shrews exist, as do aquatic and arctic forms. Shrews are rather nondescript in appearance, with short, dense fur that is usually drab gray, brown, or black. Some species have scent glands that produce a peculiar, distinctive odor, particularly during breeding season. Their eyes are small, sometimes hidden in fur, and their vision is not particularly keen, but sensitive olfactory and auditory systems make them competent hunters. Even so, a number of North American shrew species are critically endangered.

previous pages: **Star-nosed mole rears its fantastic head**

Moles, somewhat larger than shrews, tend to follow a much more underground lifestyle, for which they have developed numerous adaptations. Their eyes, like those of shrews, are reduced, in some species even covered by a layer of skin. Unlike shrews, however, the front legs and feet are highly modified for digging, and can excavate tunnels quite efficiently. The fur is dense and uniform in length, and unlike that of most mammals, can lie smoothly in any direction. This adaptation helps them back up and even turn around easily within their narrow tunnels.

Because small size and high metabolic rates go together, shrews need relatively large quantities of food. They all forage rapidly, using quick, searching movements. Some are active day and night, while others are strictly nocturnal. These highstrung little creatures will jump at the least disturbance and rarely stay still. They also make a variety of high-pitched calls, some in the ultrasonic range, which may be used in echolocation. Their sense of hearing is acute, and some species can navigate without tactile, visual, or olfactory cues. Obviously, echolocation could be a considerable advantage in locating and capturing the wide variety of small vertebrates and invertebrates that compose the typical shrew diet. Some species also produce poisonous secretions from their salivary glands that help subdue prey. They are omnivorous, in general taking more animal than vegetable food. Interestingly, they also sometimes re-ingest fecal material, much as rabbits do. This apparently helps them recapture vitamins, trace elements, and other nutrients.

Like other insectivores, moles are mostly solitary animals, getting together only to breed. Some species, however, occasionally share communal tunnels. The life cycle of most moles is rather poorly known, since their secretive nature and underground activities make them difficult to study. But in general they are active day and night, throughout the year. By pushing dirt to either side with their strong forefeet, they tunnel underground at depths ranging from just below the surface to about three feet down. When digging deeper tunnels they occasionally cut a vertical shaft, pushing out excavated dirt at the surface in "molehills." They forage for underground prey such as earthworms, grubs, and slugs in these tunnels. Some also take small amounts of vegetable matter.

Mostly solitary foragers, shrews of North America usually breed in early spring. They may produce several litters per year, with up to ten tiny, naked, blind babies born after a gestation period of about 20 to 25 days. The young grow rapidly and most species are weaned before they are a month old. The life span is probably less than a year for many species, although some individuals may overwinter and breed again the following year.

Some mole species once were harvested for their extremely soft fur, used for a variety of clothing. Moleskin has since fallen from favor, and today there is no significant trade. Seldom seen and poorly appreciated, the order Insectivora comprises a surprisingly diverse mix of creatures. Indeed, shrews and moles play important roles as scavengers and predators in a variety of ecosystems. At the same time, they also form an important prey base for larger predators. ∎

Water shrew eats a frog

Family Soricidae Shrews

Small but powerful, shrews rank among nature's tiniest predators. They fearlessly attack prey as large or larger than themselves, driven by their high metabolic rate to make long foraging bouts and satisfy their seemingly incessant appetites. Nearly constant motion and highly nervous dispositions mark most species. Some can actually die of fright following loud noises such as thunder, which can cause their heart rates to soar to 1,200 beats per minute.

Water Shrew *(Sorex palustris)* The western subspecies of the water shrew bears the stylishly appropriate name of *Sorex palustris navigator*. Distinctive looking, with a long, bicolored tail and large hind feet that bear a fringe of stiff hairs, the water shrew is a truly amphibious species. It prefers wet areas around marshes, bogs, streams, and lakes, and is often found in muskrat houses and beaver lodges. But because it also favors wooded areas, it is seldom found in wetlands that lack surrounding trees. Shunning hibernation, this little creature is active all winter, mainly at night but occasionally during daylight hours as well.

Among the most aquatic of nonmarine mammals, the water shrew can float, swim, dive, and run—both along the bottom and on the water surface. Apparently using the fringe of hairs on the hind feet to trap air, it has been seen running along the surface for some distance with its head and body entirely out of the water. This highly successful aquanaut feeds mainly on aquatic life, including a variety of insects, fish eggs, snails, leeches, even small fish or amphibians.

Pygmy shrew

Pygmy Shrew *(Sorex hoyi)* North America's smallest mammal and among the world's tiniest in terms of weight—only about 0.08 ounce—the pygmy shrew occurs solely on this continent. It ranges throughout much of Alaska and Canada, below tree line, also in parts of the Rockies and the north central and northeastern United States. It remains one of our more poorly known native small animals. Its Latin name honors American physician and naturalist Philo Romayne Hoy (1816-1892).

Recently, pygmy shrews have been found to be much more broadly tolerant of environmental conditions than originally thought. Known to inhabit wetlands and dry areas, standing forests, logged or bulldozed areas, and gardens, they also occur in close proximity to human habitation. These tiny creatures with voracious appetites feed mainly on small insects and their larvae, also earthworms, slugs, and snails. They probably fall prey to snakes, owls, small carnivores, perhaps even to members of other shrew species, with which they often share the same general area.

Family Talpidae Moles

Mammals specialized for life underground are known as fossorial animals, and moles are the quintessential fossorial mammals of North America. Some species are at least semiaquatic as well. Anatomical adaptations for their unusual lifestyle include massive shoulders and front limbs, also highly modified front feet that are used for excavating burrows and tunnels. In contrast, the pelvic girdle and hind limbs are reduced, which, along with a cylindrical body form, allows them to easily turn within the narrow confines of their tunnels. Equipped with conical snouts, tiny eyes, and small ear openings that are hidden in their fur, moles present rather whimsical faces. Curiously, their fur also lies flat in any direction, affording protection whether they are coming or going. Active throughout the year, most species feed on worms and insect larvae, although some are essentially omnivorous.

Star-Nosed Mole (*Condylura cristata*) Named for its bizarre snout full of fleshy protuberances, the star-nosed mole is among the most aquatic of moles. It is active throughout the year and also is apt to be about during both night and day, although its periods of peak activity are usually night and early morning. It swims under ice, and may be more aquatic in winter than in summer.

Unlike many other fossorial mammals, star-nosed moles are gregarious, perhaps even colonial. There are numerous records of multiple individuals inhabiting a single burrow system. Such burrows are frequently constructed near streams or bogs, and some even have underwater openings. Nests of leaves and grass are placed in

Eastern mole prefers higher, drier areas

better drained parts of the burrow, to escape flooding during periods of high water. Star-nosed moles eat aquatic worms and insects, along with a variety of other small invertebrates, including earthworms and insect larvae. Their predators include birds of prey, small carnivores, large fish, and even bullfrogs. Females are thought to produce only a single litter each year, usually in early spring, consisting of about five young, on average. One of northeastern North America's more interesting inhabitants, the star-nosed mole is rarely seen by people due to its underground and underwater habits.

Eastern Mole (*Scalopus aquaticus*) Scientific names can be both descriptive and misleading, and so it is with the eastern mole. *Scalopus* stems from Greek roots meaning "to dig" and "foot," aptly describing the highly modified forefeet of this mole. However, *aquaticus* is a misnomer, as this species much prefers upland, well-drained soils and avoids the wetlands favored by star-nosed moles. Tunnels of the eastern mole lace a variety of habitats ranging from open woodlands and fields to golf courses and cemeteries. Considered omnivorous due largely to research with captive animals, this species eats mainly invertebrates such as earthworms and larval insects that it locates in the soil. Small carnivores, birds of prey, and snakes probably are its chief predators. But eastern moles have a rather offensive odor, and this, combined with their secretive, underground habits, protects them from many enemies. Humans are undoubtedly the most serious predator of eastern moles, as mole tunnels just below the surface are unwelcome in most lawns and gardens.

Bats

Order Chiroptera Bats

BY M. BROCK FENTON

Deep inside an abandoned mine one August night in southern Ontario, the silence is broken only by the sounds of dripping water and the fluttering of bats' wings. Hundreds, maybe thousands of them are flying along the mine tunnels at speeds of 10 to 13.5 miles an hour, each one unerringly avoiding its many companions, the mine walls, and me. Most are of the same species, commonly (and appropriately) known as little brown bats. The adults have come here to mate; the young accompanying them were born about two months earlier. They begin to arrive after dark, alone or in groups of five to ten. They flit around the mine for much of the night, leaving before dawn. By the end of September, the walls and ceiling of this mine will be peppered with hibernating individuals, their numbers swelling to thousands as the season progresses.

Suddenly, one flies through an infrared beam I have set up, triggering flashes from strobe lights that freeze the bat's image—on film as well as on my retinas. The camera automatically advances film, while my eyes still "see" the bat in flight, though the mine is instantly dark again. Again and again, other bats set off the system, providing a kaleidoscope of portraits with varying postures and wing positions. The experience tends toward the surreal, particularly as I clumsily pick my way in the dark along the same passages these agile fliers negotiate so deftly.

Bats are among the most astonishing of mammals, and those of North America are no exception. While they are not the only mammals that echolocate (shrews and toothed whales also do), they are the only ones to fly. Their great longevity (perhaps up to 30 years), low reproductive rate, and intense nurturing of young— all unusual for small mammals—also set them apart. About 168 species of bats occur from southern Panama to the tree line in Alaska and Canada, compared to over 900 species worldwide. Both the largest and smallest North American species—the five-ounce Linnaeus's false vampire bat and the tenth-of-an-ounce yellow bat—are tropical. Bats of Canada, where I live, range from one-ounce hoary bats to eastern and western small-footed bats, barely larger than yellow bats.

All of them belong to the order Chiroptera, from the Greek *cheiro*, meaning hand, and *ptera*, meaning wing, since their greatly elongated arm, forearm, hand, and finger bones support wings made from folds of skin. Bats' wings attach to the sides of their bodies as well as to their hind legs, contributing to aerodynamic shape and more efficient flight. In most species, another flap of skin, the interfemoral membrane, joins the hind legs and encloses the tail.

My photographs of the cave's little brown bats showed that, although their flight seemed silent, their mouths were often open. In fact, they produce at least 20 pulses of intense, high-pitched sound per second and, between those pulses, they listen for the echoes rebounding from objects around them. Their flight seems silent to us because the sounds are beyond the range of our hearing. But bats actually are using the echoes to "read" their path ahead as they fly along. American zoologist Donald R. Griffin first coined the term "echolocation" in 1945 to describe this form of orientation and navigation.

previous pages: **Common vampire bat launches itself into the air**

All North American bats echolocate. Many use this technique to search for flying insects; greater bulldog bats, which forage over beaches and ponds, use it to detect fish by the ripples they make on the water's surface. Hunting bats generally produce intense signals, maximizing the range at which they can detect potential prey. But there are limits. Big brown bats, for example, can detect a June beetle only when it is within 15 or 16 feet. Even the most intense of echolocation signals do not give bats nearly the range of detection that we achieve with vision.

The impressive diversity of bats stems in part from the different ways they find their food. Gleaning bats—those that take insects and other prey from the ground or foliage rather than plucking them from the air—are far less dependent upon echolocation to find their food than are bats that hunt for flying insects. Some gleaners act much like barn owls, locating prey by the sounds they make as they move. Others rely on rather keen eyesight—for, no, bats are not blind. Gleaning bats produce much softer echolocation calls than aerial-feeding species, as do bats that eat fruit, nectar, pollen, or blood. While most North American bats, such as little brown bats, emit echolocation signals through open mouths, New World leaf-nosed bats (family Phyllostomidae) emit signals through their nostrils. As their name implies, leaflike structures dominate the faces of most such species.

Compared to most other mammals, bats are tiny. The largest bat species in the world, the biggest flying foxes, can have six-foot wingspans—but weigh less than four pounds. Those that hunt airborne insects are far smaller, most weighing in at little more than an ounce when fully grown. Only four or five aerial-feeding species exceed three ounces. Perhaps their relatively short-range echolocation system, combined with the size of their insect prey, helps explain why these hunters are so small. For while a bat-size animal can tolerate a short range of operation, a larger animal cannot.

Yet in spite of their diminutive size, bats are long-lived. North America's current record-holder is a little brown bat last seen in the wild in southeastern Ontario, 33 years after it had originally been banded as an adult. While we do not have comparable band records for most other North American bats, it appears that many commonly reach the 10-to-20-year range, an exceptional span for such tiny creatures. Bats also differ from the small-mammal norm in that they have relatively low rates of reproduction. Little brown bats, for example, bear one young per year, while big brown bats and hoary bats more often have twins. Some tropical North American species give birth to a single young twice a year.

Most bats mate at the end of summer, when they are in good condition and food is still abundant. Ideally, the young are born no earlier than the following spring, when food is again plentiful. Such a scenario, however, could mean a very long pregnancy, long especially for such tiny mammals. Bats sidestep this problem in either of two ways. Some, such as California leaf-nosed bats, immediately become pregnant after a successful mating; but the development of the young is delayed for much of the winter and resumed only in spring, when the pregnancy goes to term. Plain-nosed bats take a different approach; although they also mate in early fall, females of these species store sperm in their uteruses until spring, when ovulation and fertilization occur. Some of them can retain viable sperm for up to 200 days, longer than males store them inside their own bodies!

Baby bats are enormous compared to adults, as newborns often weigh about one-fourth as much as their mothers—the equivalent of a 130-pound woman having a 32-pound baby. Like all female mammals, mother bats feed their young milk produced in mammary glands, located on the chest. Bat milk is extremely rich, and the young consume a great deal, often taking in their own weight in milk every day. This presents two challenges to the females, particularly to those living in colonies. First, they must eat enough food to produce their highly nutritious milk. Some lactating bats eat their own body weight in food every night—a major accomplishment. But to be successful mothers, they also must be able to consistently find their own youngster amid a sea of others, whenever they return from feeding forays. Easily the most gregarious of mammals, bats congregate in colonies that can number in the millions. How does a mother find her baby? Excellent spatial memory enables her to locate the general area, then she homes in on her youngster's specific vocalizations and scent. Baby bats, like the young of many species, often try to get milk from any female. But the mothers are far more discriminating, usually allowing access only to their own young. Thus nature ensures that the mother's huge investment in the next generation will pay off.

Roosts are vital resources for bats, whether they consist of nursery colonies, which harbor females and young, or all-male bachelor colonies. Bats roost in an astonishing variety of situations. In tropical North America, several species of New World leaf-nosed bats roost in tents they make from leaves. Other species commonly roost among the foliage of trees and vines, relying upon their small size and inactivity to escape the attention of predators. Both the tent-makers and the foliage-roosters, however, are largely exposed to ambient temperatures. Many other species prefer the shelter of hollows or cracks or crevices, which protect them from temperature extremes as well as from predators. Large colonies can generate considerable collective body heat, offsetting the effects of a cool day. Some species enter torpor—a kind of cold-induced sleep, similar to hibernation but much shorter—when temperatures fall. Torpor allows a bat's body temperature to approximate that of its surroundings. Of course, activity is also curtailed. Torpor is an energy-saving behavior that is key to the survival of bats that use it.

Bats that inhabit North America's colder regions generally eat insects and other arthropods, prey that become much less available during winter. Consequently many of them hibernate, often seeking out caves, abandoned mines, and other underground locations where they benefit from above-freezing temperatures and relatively high humidity. Some hibernators, such as little brown bats, may migrate a few hundred miles from summer to winter homes. Other species, notably red, hoary, and silver-haired bats, make more extensive migrations, summering in northern Canada and wintering as far south as the southern United States. They seem to roost in rather exposed locations, a strategy that works only where cold snaps are brief and temperatures do not fall far below freezing. Like plain-nosed bats, these species also may use torpor to pass short periods of inclement weather. Some of the longest migrations are performed by Mexican free-tailed bats and Sanborn's long-tongued bats, both of which leave the southwestern U.S. to winter in Mexico.

Today, about half a dozen North American bat species have been listed as endangered, due to dramatic declines in their populations. Like all creatures, bats have certain needs that must be met if they are to survive. They need secure roosts, the right climatic conditions, and of course food. Conservation of bats is good not only for them but also for us. It's been estimated that 100 little brown bats can eat about 7 pounds of insects a night, or 840 pounds over a 120-day-long summer. If you think the midges and mosquitoes are bad now, just imagine how infernal they could be if their chief predators, bats, were removed from the ecological equation.

Biologists have found that, since different bats play different roles—some eating insects, others fruit or nectar or pollen—monitoring their populations can help assess the planet's overall health. The disappearance of a certain bat species from one area may indicate loss of food sources—itself a signal of ecological change— or perhaps climatic shifts. Bats make especially good indicators of long-term changes, for they are long-lived, small in size, and highly mobile. Bats also possess what biologists call behavioral flexibility, in that they often exploit human-caused environmental changes to their own advantage. One outstanding example is the common vampire bat, whose populations appear to have expanded in response to increases of domestic livestock. In addition, some North American bats readily roost in abandoned buildings and mines; others exploit the swarms of insects that congregate around streetlights.

I consider bats to be the most astonishing of mammals. They have kept me enthralled for over 30 years, always presenting new challenges and the promise of new discoveries. Whether the topic is echolocation or roosting behavior, bats remain mysterious, full of unknowns. Their eyes are a prominent feature of most species—yet many people incorrectly think of them as blind. In fact, I fully expect that one of the most exciting areas for future bat research will concern vision and the role it plays in the daily lives of these fascinating creatures.

But perhaps the biggest challenge will continue to be overcoming their negative image. As creatures of the night, bats have been viewed with suspicion for much of history. They have been associated with disease, such as rabies. And, although biologists find the many specializations of vampire bats captivating, many people are repulsed by the idea of an animal that feasts on blood. Most bats, of course, do not. Yet we have allowed the three species that do to affect our perception of all others.

Bats are just the animals to study if you believe that science has all the answers; there is so much to be learned about them, and we have only just begun. ■

Big Brown Bat

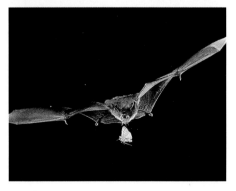

Spotted Bat

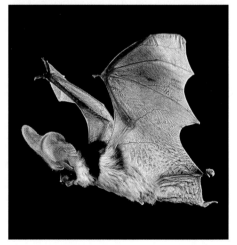

Greater Funnel-Eared Bat

California Leaf-Nosed Bat

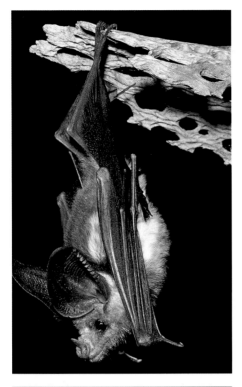

Family Vespertilionidae   Vespertilionid Bats

Big Brown Bat *(Eptesicus fuscus)* [page 32, left column] Poised for take-off, this big brown bat scans its world using both vision and echolocation, hence the open mouth. It hunts flying insects the same way, with eyes and mouth wide open. These bats weigh less than an ounce and typically live in colonies of about fifty to several hundred individuals. They commonly roost in buildings throughout eastern North America, often returning to the same place day after day and year after year. In southern British Columbia, however, they most often roost in hollow trees, changing locations several times a week. Rabies is rare in all bats, but this species is the one most often tested for it. Its habit of roosting in buildings often puts it in contact with humans. Like many other wild animals, bats bite in self-defense; a sure way to avoid such bites is to leave them alone.

Spotted Bat *(Euderma maculatum)* [page 32, right column] Enormous, swept-back pink ears and black fur with white spots are among the most striking features of this insectivorous species. Pleats permit the ears to fold back easily. Spotted bats weigh about the same as big browns, and range from northern Mexico through the western United States into southern British Columbia. Radio tracking studies show that they roost by day in the nooks and crannies of cliff faces; they are not known to roost on buildings. Nor do they seek large concentrations of insect prey, instead flying and foraging all night long. Unlike most other North American bats, they use echolocation calls that are within the range of human hearing. They often eat moths, many species of which hear higher frequency echolocation calls but do not hear the calls of spotted bats until it is too late. We still do not know the functional significance of the white spots on their backs.

Family Natalidae   Funnel-Eared Bats

Greater Funnel-Eared Bat *(Natalus stramineus)* [page 33, left column] Weighing in at only a tenth to a fifth of an ounce, this delicate insect eater has two color phases. It usually roosts in caves, at times in very large numbers; females bear a single young each year. Funnel-eared bats are relatively common, but we know little about their behavior. Four other species complete this family, which occurs in the West Indies and Central America, as well as in South America.

Family Phyllostomidae   New World Leaf-Nosed Bats

California Leaf-Nosed Bat *(Macrotus californicus)* [Page 33, right column] A projecting nose, large eyes, and striking ears dominate the face of this southwestern species. The projection, called a noseleaf, directs echolocation calls as they are emitted through the bat's nostrils. The large eyes speak to visual abilities that, at night, exceed our own. The large ears are particularly sensitive to low-frequency sounds such as the faint footfalls of walking insects. California leaf-nosed bats appear to use both their acute vision and hearing to detect, locate, and assess potential targets, but we know little about the role echolocation plays in their lives. They frequent arid habitats from southern California and Arizona into Mexico, typically roosting in caves and abandoned mines and often preferring

geothermal areas where underground heat helps them survive winter temperatures. They are one of 76 species of leaf-nosed bats in North America.

Pygmy Fruit Bat (*Artibeus phaeotis*) [above] Clinging to its meal with its thumbs, a pygmy fruit bat devours a piece of banana. This species occurs from southern Mexico through Central America to northeastern South America, often roosting by day in group "tents" constructed from leaves it modifies by selectively biting through the leaf veins. A typical tent includes one adult male and several females, as well as their dependent young. Female pygmy fruit bats bear one young at a time but often produce two offspring a year. They feed mainly on fruit, also taking nectar and pollen from time to time. They occupy a wide range of habitats, from tropical deciduous forest to thorn forest.

Common vampire bat laps blood from a victim

Common Vampire Bat *(Desmodus rotundus)* Among the most maligned of creatures, vampire bats are the only mammals known to feed solely on blood. The common vampire bat uses razor-sharp upper incisor teeth to remove a tiny divot of skin from its sleeping host; anti-clotting chemicals in its saliva and an active tongue promote bleeding, enabling the bat to drink its fill—usually about two tablespoons. Its acute sense of hearing detects the breathing sounds of sleeping prey, while infrared sensors on its face help target areas of skin where blood flows near the surface. Even so, it fails to feed about one night a month; youngsters may miss one night in three—a major risk, since they cannot survive two nights in a row without eating. Vampire bats that return to a roost with full stomachs often regurgitate some blood for their roost mates, thus providing a social safety net.

Western mastiff bat

Family Molossidae Free-Tailed Bats

Western Mastiff Bat *(Eumops perotis)* Largest bat in the U.S., the western mastiff weighs from 1.8 to 2.5 ounces. It is insectivorous, occuring in two populations—one from southern California, Arizona, New Mexico, and Texas south into Mexico, the other in South America. Like all free-tailed bats, it has a thick tail extending beyond the membrane between the hind legs. Its large ears, joined at the middle of the top of the head, extend beyond the tip of the nose. Females generally produce one young yearly, establishing nursery colonies in cracks in cliff faces or buildings, places that ensure them sufficient free fall to build airspeed, since these bats cannot take off from the ground. Their long and narrow wings are made for rapid, economical flight, while their short, silky fur minimizes drag. Their low-frequency echolocation calls, like those of spotted bats, make them relatively inaudible to moths in spite of their large size.

Family Mormoopidae Mustached Bats

Named for the mustache-like whiskers and thickened flaps of skin around their mouths, mustached bats include eight species worldwide; five occur in North America. Only one, the ghost-faced bat, inhabits the United States, specifically parts of Texas near the Mexican border. It weighs up to an ounce in size, usually roosts in caves, is insectivorous, and uses echolocation to search for prey. Images of ghost-faced bats have been found on some Native American pottery, both in Central and South America.

BY DON E. WILSON

Once called Edentata, meaning toothless, the mammalian order that includes armadillos, sloths, and New World anteaters has since been christened Xenarthra, meaning "strange joint." The new name refers to a vertebral arrangement that helps its members dig more efficiently. "Edentata" was always a misnomer, for among all the species of this group, only anteaters lack teeth. The lumbar reinforcements dubbed "xenarthral" provide a far more accurate signature for the order, differentiating it from all other mammal groupings.

By any name, this exclusively New World order makes an interesting collection, with a rich and varied history. Extinct xenarthrans include the huge, armored, armadillo-like glyptodonts as well as ground sloths—large terrestrial relatives of modern tree sloths. In fact, the order once was remarkably more diverse than it is now; it boasts more than ten times as many fossil genera as current ones (which number just 13). Like marsupials, xenarthrans were strongly influenced by the long separation of North and South America during the Tertiary period. Much of their early evolution occurred in South America, and by Pliocene times (a few million years ago) the land bridge along the isthmus of Panama allowed giant ground sloths to move northward. Glyptodonts and armadillos followed later.

Today's tree sloths occur only as far north as northern Central America, anteaters as far as southern Mexico. Only a single armadillo species—*Dasypus novemcinctus*, the nine-banded armadillo—managed to succeed in the United States. Like the Virginia opossum, it is a curious and hearty survivor. First recorded north of Mexico in 1849, this miniature tank of an animal has steadily increased its range northwards and eastwards from its initial foray into the state of Texas.

Armadillos currently roam from Kansas across the southeastern U.S. to South Carolina. While individuals have been found well beyond this range, even in Delaware and Washington, D.C., they almost surely represent introduced animals. East Coast populations are an extension of introductions in Florida during the 1920s, after several individuals escaped from small zoos and private owners. The armadillo's northern limits are determined by winter temperatures in unusually cold years, and global warming may be extending those limits, encouraging the continued advance of warm-weather creatures such as armadillos and opossums.

Even the name armadillo, Spanish in origin, reflects this animal's southern roots. It means little armored one, an apt description. The "armor" consists of epidermis modified into a series of large horny scales, connected together in bands or plates that are separated by more flexible skin. Such construction provides the sole North American species, the nine-banded armadillo, with considerable protection from predators. The animal's underside and portions of its limbs have soft, hairy skin; during times of threat, these vulnerable areas are pulled under the hard dorsal carapace, much as a turtle's body retreats into its "shell." Although nine-banded armadillos do not curl into a ball, they will wedge themselves into a burrow or crevice, their rounded carapace protecting the underparts and making the animal extremely difficult to dislodge.

Prime defense posture for a nine-banded armadillo

Xenarthrans

Armadillos are basically forest dwellers, but their northerly populations have expanded into habitats much more arid and open than traditional forests. Although appearing sluggish, they actually are capable of impressive leaps and quick bursts of speed. They also are accomplished burrowers, constructing underground nests lined with leaves and grass.

Most active at night, armadillos occasionally move about in the daytime during cool weather. They forage busily, poking through leaf litter and into the ground for arthropods, small vertebrates, and occasional fruit or vegetable supplements. Their expansion within the U.S. has been marked by a tendency to amble across highways. This and their normal defense posture—leaping up to scare off a predator—have, unfortunately, made them frequent victims of automobiles.

Nine-banded armadillos have an unusual reproductive cycle: Each fertilized egg gives rise to genetically identical quadruplets. This unusual trait makes the armadillo a useful animal for genetic studies, since experiments can be carried out on genetically identical animals. Another reason to study this creature: Apart from humans, it is the only animal known to be susceptible to leprosy.

Individuals reach sexual maturity in about a year, and mating usually occurs by midsummer. Implantation of the fertilized egg, however, is delayed until November, enabling the young to be born in late winter or early spring rather than in a more challenging time of year, which would occur if the egg implanted at the time of fertilization.

Armadillos show no signs of territoriality. Although individual home ranges may extend over 50 acres, these solitary creatures occur in population densities as high as one or two animals per acre. Individuals generally display little aggression toward each other despite such overlaps. Each constructs multiple burrows, usually between four and eight, within its home range. While their burrowing activities may be disruptive to agriculture in some areas, on the whole they are probably beneficial, since they eat harmful insects.

One curious behavior reported for armadillos is their ability to walk underwater. This is most likely in shallow streams, as they can remain below the surface for only several minutes. Their specific gravity—the ratio of body density to the density of water—allows them to sink. Armadillos also are capable swimmers; apparently they can increase their buoyancy by gulping air into their digestive tract, allowing them to ride fairly high in the water. In this position they swim much like other four-legged mammals, by paddling with their legs.

The entire order Xenarthra, organized into separate families of anteaters, sloths, and armadillos, is another successful group of mammals with ancient roots. Each family has specializations for different and quite distinctive lifestyles. Anteaters feed mostly on social insects such as ants and termites; sloths browse their way slowly through tropical forests. Armadillos are much more general, both in habitat requirements and food preferences, and this flexibility has allowed them to extend their range much farther north than anteaters or sloths. ■

Family Dasypodidae Armadillos

Nine-Banded Armadillo *(Dasypus novemcinctus)* Sole member of this biological order to occur north of Mexico, the nine-banded armadillo is the unmistakable armored car of our small-mammal world. In contrast to the horny outer scutes that cover most of this creature's upper surfaces, its snout tapers to a piglike softness. Long, hairless ears add to the curious external appearance. About two-and-a-half feet long and weighing 12 to 17 pounds, armadillos are surprisingly light on their feet. Their common reaction to a threat is to immediately leap high in the air at a single bound *(page 38)* and then to hit the ground running.

Cold winter temperatures restrict these tropical immigrants to the southern half of the United States, and their continued expansion may become a useful indicator of global warming in the coming millennium. Armadillos occur in a variety of habitats, most often in areas with loose soils, for they are burrowers by nature, relying on powerful legs, claws, and a good sense of smell to root for prey. Primarily nocturnal, they sometimes begin foraging by late afternoon. Small invertebrates such as insects are their main food, with reptiles, amphibians, and occasional vegetable matter rounding out the diet. Most prey items are relatively soft, since the tiny, peglike teeth of armadillos cannot easily crush hard foods.

Lagomorphs

Order Lagomorpha

BY ANDREW T. SMITH

Forty years ago a respected scientific publication carried an article entitled "What, if anything, is a rabbit?"

Today this question strikes us as odd; doesn't everybody know what a rabbit is? Yet in its day it pointed up a long-standing controversy as to whether or not rabbits should be lumped together in the same order as rodents. After all, both forms possess large front incisors and are primarily herbivorous. But by the time the dust from this debate cleared, rabbits were considered independent. We now know that the fossil history of rabbits extends back about 60 million years, to a time when many other major forms of mammals became distinct.

What I have liberally referred to as rabbits are actually members of the order Lagomorpha, which includes two lesser groups: pikas (family Ochotonidae) and rabbits and hares (family Leporidae). "Rabbit" and "hare" convey little scientific meaning, although the first term normally refers to smaller species such as cottontails and the second applies to larger ones such as jackrabbits. Lagomorph simply means hare-shaped; all members of the group have very short tails and hind feet at least somewhat larger than forefeet. While the hind feet of pikas are only marginally bigger, those of rabbits and hares can be considerably so.

Lagomorphs also are known for their ears, but again, size can vary greatly. Jackrabbits sport gargantuan ears reminiscent of dish antennas. Ears of cottontail rabbits are slightly less pronounced, while pikas have rounded ears, relatively wide for their head size. In addition to ears and feet, another clear-cut feature sets lagomorphs apart from rodents: their teeth. While both groups have large, continuously growing incisors, only lagomorphs possess a second set of small, peglike incisors that lie just behind the large ones. No one knows the functional importance of these teeth, but they remain a distinguishing trait.

Pikas, also known variously as conies, rock rabbits, whistling hares, and little chief hares, occur in our continent in two species, the American pika (*Ochotona princeps*) and the collared pika (*O. collaris*). Both inhabit the western mountains, with collared pikas existing only in Alaska and northwestern Canada. Both are denizens of high alpine talus. Active in daytime, these smallest of lagomorphs carve out individual territories amid the loose rock adjoining alpine meadows. Like all members of the order, they do not hibernate. Instead, they construct a hay pile— a cache of food—to carry them through the rigors of winter. Perhaps the most familiar antics of pikas during summer are their endlessly repeated forays to alpine meadows to mow down plants and carry them back to add to their hay pile.

Pikas are the most vocal of all lagomorphs. Only males give the long call, which sounds like a song and is given primarily during the breeding season. Both sexes utter the short call (a plaintive "eeek"), to proclaim territory and to warn of predators. Pikas also spend much of the day in quiet repose, generally sitting high atop prominent, backward-sloping boulders. One of my favorite pastimes is to mimic a sitting pika, taking in all the beauty of the alpine summer while watching the neighbors go about their business.

previous pages: **Black-tailed jackrabbit**

North America's native rabbits and hares have been classified into four genera. Mexico's volcano rabbit (*Romerolagus diazi*), also called *zacatuche* or *teporingo*, inhabits only the high volcanoes surrounding Mexico City, and is considered one of the most primitive of lagomorphs. The pygmy rabbit (*Brachylagus idahoensis*) is the smallest of the group, thriving amid sagebrush of the northern Great Basin. Cottontails (genus *Sylvilagus*) are the most diverse, numbering 14 different species; they occur only in the New World. Hares and jackrabbits (genus *Lepus*) are the largest of lagomorphs and are broadly distributed throughout the world.

Two lagomorph species have been introduced into North America. One, the European rabbit (*Oryctolagus cuniculus*) now lives in the San Juan Islands of Washington and the Farallon Islands of California, where it occurs in such numbers that it is considered a pest. The same species invaded Australia in 1858, causing rangeland damage and contributing to the extinction of many native marsupials. The European hare (*L. europaeus*) has expanded throughout the northeastern U.S., where it was first introduced as a game animal.

Like all herbivores, lagomorphs face a basic problem: How to digest cellulose when they, like all other animals, are unable to synthesize the enzyme that digests this most basic component of plant tissue. They solve this problem in a particularly distinctive if unappetizing way. In addition to solid feces, lagomorphs produce a second kind of dropping—soft black viscous pellets resembling axle grease— which they swallow as soon as they eliminate them. The reason for this has to do with their peculiar digestive system; among lagomorphs, chewed plant material normally collects in the cecum, a secondary chamber between the large and small intestine. This chamber contains symbiotic bacteria, which aid in the digestion of cellulose and also produce certain B vitamins. Soft feces form here, containing up to five times as many vitamins as hard feces. After they are excreted and then eaten, the soft feces are redigested in a special part of the stomach. This double-digestion process enables lagomorphs to utilize nutrients they may have missed the first time, thus ensuring that they get maximum nutrition out of the food they eat. In fact, deprived of the opportunity to ingest soft feces, some species may die.

We've all heard the expression "to breed like a rabbit," and indeed, many rabbits and hares produce large numbers of young per year. Several factors help make this possible: First, some species regularly conceive litters of up to six or seven young, often four or five times a year. Second, most rabbits and hares reach sexual maturity at a relatively early age, some females becoming reproductively active only a few months after they are born. Third, lagomorphs exhibit induced ovulation, their ovaries releasing eggs in response to copulation rather than on a regular cyclic basis. Fourth, females can undergo a postpartum estrus, which allows them to conceive immediately after giving birth. The combined effect of the last three factors is to minimize the interval between births; the result is similar to that of a compound interest problem—the sooner and more often an animal can breed, the greater its potential to produce many young.

Not all lagomorphs, however, are highly fecund. In the far north, where the summer breeding season is short, Alaskan (*Lepus othus*) and arctic (*L. arcticus*) hares generally conceive only one large litter of about six young per year. In contrast, black-tailed jackrabbits (*L. californicus*) and antelope jackrabbits (*L. alleni*), which inhabit more southerly and drier habitats, face a scarcity of resources over an extended breeding season. They produce four or five small litters a year, each usually consisting of one to three young.

Pikas further demonstrate how ecological realities can influence reproduction. Both species of North American pika generally wean only two young per year. Although they normally start with three embryos, females usually either resorb at least one fetus before giving birth or abandon one baby while nursing—apparently due to the difficulties they face acquiring sufficient resources to raise their young. It is also significant that, although female pikas mate twice each season, normally only the first litter survives—presumably because mothers cannot ordinarily muster enough energy to raise two. The second litter appears to be a sort of insurance policy, should the first litter fail for any reason.

One of the most remarkable aspects of lagomorph reproduction is how terribly inattentive the mothers are to their young—and how they get away with this. They are very nearly absentee parents; rabbit and hare mothers commonly nurse their young only once every 24 hours. Some researchers have observed mothers that did not return to their nests for over 30 hours! In addition, the duration of these visits is quite short, allowing little time for transfer of milk to the young. Yet the young survive because the milk is highly nutritious, in fact among the richest of all mammals. Pikas appear to be this order's most attentive mothers, visiting their young as frequently as every two hours. In spite of the apparent lack of maternal care, young lagomorphs grow rapidly and are weaned in about a month—the approximate age of independence for most mammals of this size.

About 80 lagomorph species exist worldwide (27 of which are native to North America), compared to more than 2,000 species of rodent. Why so few lagomorphs relative to rodents? We don't know. But we do know that what lagomorphs lack in number of species they make up in the wide range of ecosystems they occupy throughout North America. From the Arctic north to the deserts of the Southwest, from northeastern deciduous forests to midwestern grasslands, from alpine regions to the tropics of Mexico, lagomorphs are found nearly everywhere. In the United States, sport hunters take more rabbits than any other game animal, avian or mammalian. Cottontails are particularly important to eastern and midwestern states where other game is relatively scarce. The high fecundity of many of our lagomorph species also translates to their importance as a food source for many of the predators of North America, thus highlighting the important role they play in natural food chains.

Snowshoe hares (*L. americanus*) of the boreal forest of Canada exhibit extraordinary variability in numbers. In some years they abound; in other years one is hard-pressed to find any. Their populations go through a ten-year cycle of boom and bust, and at times it seems that only their boom-time populations exceed the number of theories posed to explain this enigmatic and regular phenomenon. Most contemporary research indicates that predation plays a large role in the ten-year cycle, but many subtle nuances of behavior and ecology also enter the equation. Predator populations build up as rabbit populations increase, eventually creating so many predators that the hares take a nosedive. Then, after the predators starve for lack of hares, the prey rebound.

Recent behavioral research has shown another effect of high predator density—those hares that manage to survive the initial onslaught of predation become increasingly wary. They minimize their time foraging on preferred foods, which often grow in open clearings where they are most likely to be nailed by predators. Forced to eat less nutritious foods, they experience declines in overall condition and eventually in their normal reproductive rates. Predation risk is not constant during the ten-year cycle, and whenever the predators die off, snowshoe hares again venture into the open. They soon fatten up, reproduce well, and their numbers rebound, thus completing the cycle.

Based on our understanding of cottontails and snowshoe hares, it is easy to assume that all lagomorph species should be common. Unfortunately this is untrue. Some of the most endangered mammals in the world are lagomorphs, and several of these occur in North America. Mexico harbors 15 lagomorph species, 8 of which are endemic to the country; 4 are listed by the IUCN as critically endangered or endangered; 4 more are listed as near-threatened.

One, the Omiltemi rabbit (*S. insonus*), is known to occupy a small region in the mountains of Sierra Madre del Sur in Guerrero, where only three individuals have been collected since the original description of the species in 1904. Recent comprehensive surveys have been unable to locate this rabbit, and we are very close to calling it extinct.

Another Mexican endemic, the small and primitive volcano rabbit, has the misfortune to live only in *zacaton* grass habitat in the highland pine forests on the volcanoes surrounding Mexico City. We have proposed that this species become the conservation symbol of Mexico. It is indeed an important indicator, reflecting the overall health of an ecosystem that includes over 370 endemic species of plants and animals. But unsustainable hunting, pollution, and encroachment on its habitat by the continued growth of the world's largest city all conspire against this rare rabbit. It is sometimes difficult to convey to people that something as notoriously prolific as a rabbit can be endangered, but the truth is that all the cards are stacked against the volcano rabbit.

So, what is a rabbit? We have seen that rabbits, hares, and pikas form an interesting, distinctive, and dynamic alternative to rodents among small mammals. The importance of many lagomorph species in our ecosystems, as well as the conservation needs of many others, make them worthy of our attention and study. ■

Family Ochotonidae **Pikas**

American Pika *(Ochotona princeps)* Pikas are at home among the rocks. These exclusively herbivorous animals inhabit and defend territories that consist solely of talus. Frequent trips to adjoining alpine meadows are essential, providing both fresh forage and a source of hay the animals gather and store in order to overwinter. Such trips, however, expose them to greater risks of predation. Amid the talus, pikas can instantly hide. But off the rocks, predators can easily run them down. Thus pikas carefully limit their time in the meadows. They are prudent shoppers, gathering those plants that best serve their purposes.

Whenever pikas leave the safety of the talus to forage, they first must decide whether to feed or to gather hay. Feeding pikas normally venture only a yard or two from the talus, to well-groomed areas where continuously growing grasses offer a constant source of nutrition. Because these closely cropped areas afford great visibility, they also serve as veritable observation posts from which pikas can scan for approaching predators.

When gathering hay, however, the animals roam farther and farther from the talus, searching for plants large enough to make each harvesting trip worthwhile. They cram their mouths full, choosing plants that are highest in protein and most likely to remain fresh over time. They return, perhaps trailing yard-long stalks of bluebells. Their hay piles help sustain them throughout the long winter as they hunker down under the snow and await another glorious alpine summer.

Pika returns with fresh cuttings for its hay pile

Family Leporidae Hares and Rabbits

Arctic Hare *(Lepus arcticus)* Resident of the Far North, this most unusual lagomorph remains a creature of the snows. Its range includes Ellesmere Island and northern Greenland, where they remain white year-round. Most populations shed their white winter coats for gray or light brown summer pelage, thus affording themselves appropriate camouflage for each season, to help them escape predators. When not looking for food, arctic hares may crouch for hours amid lichens and twigs of other hardy alpine plants, careless of the dry drifting snows that can completely bury them as they sleep. Although most hares are extremely unsocial, this species is positively gregarious. Herds of a hundred or more animals have been seen dotting some windswept tundra uplands. Occasionally, individuals will hop upright like kangaroos; in spring, males assume a similar stance [above] while fighting each other for mating opportunities.

Black-Tailed Jackrabbit *(Lepus californicus)* Bouncing west by stagecoach, Mark Twain encountered a black-tailed jackrabbit. *Roughing It* chronicles the occasion: "As the sun was going down, we saw the first specimen of an animal known familiarly over two thousand miles of mountain and desert— from Kansas clear to the Pacific Ocean—as the 'jackass rabbit.' He is well named. He is just like any other rabbit, except that he is from one-third to twice as large, has longer legs in proportion to his size, and has the most preposterous ears that ever were mounted on any creature but a jackass."

The name—shortened to "jackrabbit"—stuck. Enormous ears not only help these desert dwellers detect approaching predators but also help dissipate heat. Hold a jackrabbit's ear to the sunlight and you see a rich supply of veins; the ear's large surface area and extreme thinness make it an ideal organ for cooling the blood. Little wonder that more northerly species of rabbits and hares have shorter ears, better suited to heat retention than to heat loss. The jackrabbit's long legs make it a racer; even when being overtaken by a pursuer, this species relies on sheer running ability rather than dodging into thick brush or diving into a burrow.

Black-tailed jackrabbit preens

Rodents

Rodents

Order Rodentia

BY NUMI C. MITCHELL

I waded through the Ocracoke salt marsh, my eyes fixed on the ground a few feet ahead; I had come to North Carolina's Outer Banks in search of the Outer Banks king snake, a mysterious subspecies of the mainland king snake. Two of them had been discovered some 30 years earlier; ever since, people had tried to find others, without success. They had searched mostly in upland forests, where mainland kingsnakes live. But just yesterday one of our team had captured the world's third Outer Banks king snake—in a salt marsh—and I wanted to catch one, too.

A large piece of plywood, cast up by storm tides to the wavering line of wrack that marked the marsh's inland edge, caught my eye. I felt my heartbeat rise; such a sizeable item might shelter a snake. I approached cautiously and quietly, then quickly flipped the board. No king snake. But something furry perched in the center of a cup-shaped nest. I pounced on it with a flat hand.

When I first held it up, I thought it a mouse. But it was too big. The fur was woolly, the eyes were large, the ears small; its face had a bowed profile, while the bicolored tail was only sparsely furred. It struck me as a Beatrix Potter sort of rodent; in fact, it was my first rice rat.

I put it into my snake bag and examined the nest, built above the water and supported by cut stems of grass that acted just like boardwalk pilings. Then I noticed runways, also elevated, leading out from the nest. I flaked the top layer off a pile of dead rushes near the plywood and saw that the entire wrack line was mazed with similar rat trails. Suddenly I realized my bag held the answer to the king snake mystery. I knew that rice rats were the only native rodents to inhabit vast stretches of the Outer Banks, because only they can survive the winter flooding. King snakes prey primarily on rodents, so the presence of rice rats in the salt marsh must explain why the snakes were here. I started asking myself what physical adaptations and ingenious strategies this rat might have that enable it to survive where no other rodents can. For me, this began what has become an enduring fascination with rice rats, and with rodents in general.

Rodents are the most varied order of North American mammals, both in terms of their habitats and their physical adaptations. They are also the largest order, in number of species as well as in total population. More than 200 rodent species inhabit the U.S. and Canada—nearly as many as all other mammals combined. They are primarily herbivorous, harvesting seeds and all kinds of plant tissues, from roots to branch tips. But the most defining characteristic of this group concerns teeth: All rodents are equipped with paired, ever growing, chisel-like incisors with which they grasp, hold, cut, pierce, and—most importantly—gnaw food. Indeed, the name of the order stems from the Latin *rodere*, "to gnaw."

While dentition and diet unify rodents, as a group they are strikingly diverse. Some have developed bipedal locomotion; others possess flattened tails and webbed feet for swimming, spines for defense, skin flaps for gliding, prehensile tails for climbing, scoop-like forepaws for digging, and a range of physiological and behavioral specializations that suit them to a wealth of different habitats.

previous pages: **Black-tailed prairie dogs munch and watch**

Much of the diversity we see in rodents has been dictated by competition, both from within the order and from without: Herbivorous insects, birds, rabbits, pikas, ungulates, and others all compete with rodents for a finite amount of plant matter. The rodent strategy has been to specialize; most rodent species are very adept at exploiting only one or two layers of the environment. They can find food and shelter in these layers more easily and efficiently than generalists, so they survive—as long as their habitat does. Most rodents have developed specializations for either arboreal, terrestrial, aquatic, or fossorial (underground) habitats.

Predation also has been significant in shaping rodent evolution. Rodents are an important food source for numerous predators higher up the food web: snakes, raptors (hawks and owls), canids (foxes, coyotes, and wolves), cats (bobcats, lynx, panthers, and ocelots), and mustelids (weasels, minks, otters, and such). With such broad predation pressure and with few defensive weapons apart from their teeth, most rodents—not surprisingly—have evolved the vital quality of being fecund. Their common strategy, like that of lagomorphs, seems to be to saturate their environment with many ill-cared-for offspring rather than to invest much time in rearing a few, well-protected young; they are gambling that at least two will survive to replace the parents and reproduce.

Through the coupled pressures of competition and predation, rodents have evolved numerous mechanisms to sense threats and avoid being eaten, while exploiting their environment. Typical body modifications involve placement and size of the eyes and relative tail length. Species that forage in exposed areas and trees have large eyes located on the side of the head, for maximum peripheral vision, and elongated tails that give them balance, stability, or the ability to steer themselves. Tail development is most extreme in arboreal and flying squirrels, which use their long furry tails as rudders to correct their trajectory during leaps. Golden mice and rice rats instead use their tails like a fifth limb, wrapping it around branches as they lean out to pluck hard-to-reach seeds or fruit.

On the ground, kangaroo rats and jumping mice manage to walk and hop bipedally, a trait that permits speedy and unpredictable movement during escapes. The tail is used both as a counterweight to prevent somersaulting during a leap, and as a prop when the animal stands.

Fossorial animals, which live underground or use tunnels through soil or grass, tend to have reduced eyes placed toward the top of a flattened head, also shorter tails and smaller ears than surface-dwelling species. Eyes are not much use underground, but they can be very important when the animal emerges from a burrow or is uncovered by a digging predator. Stability is not an issue in a narrow tunnel, so there is little need for an oversize tail; large ears also just get in the way. Instead, fossorial animals often possess sensitive vibrissae—whiskers—and various pressure receptors that warn of approaching predators. Such rodents include (from most fossorial to least) pocket gophers, mountain beavers, woodchucks, lemmings, voles, and cotton rats.

Order Rodentia **Rodents**

Then there are the rodents that maintain extensive burrow systems but forage above ground: grasshopper mice, prairie dogs, chipmunks, and some ground squirrels. These species display both fossorial and terrestrial characteristics: large eyes, relatively short tails, and—except for grasshopper mice—small ears.

Most rodents can swim if pressed, but a few—beavers, muskrats, and nutrias (introduced from South America)—are highly specialized for aquatic existence, with paddle-like feet and oar-like tails. The eyes, however, seem fossorial: small and located high up the head. In fact, they ride just above the waterline when the animals swim, which allows them to keep watch for approaching predators.

Rodents generally tend to be either crepuscular (active at dawn or dusk) or nocturnal (active at night), because they are less visible to predators in the dark. Large, protruding eyes maximize the light-gathering ability of such species. Large ears serve as hearing trumpets, magnifying the sounds around them and giving them an edge in detecting predators. Kangaroo rats and pocket mice even have enlarged auditory bullae, bony resonating chambers, inside their ears. Many rats and mice are equipped with a ring of thin skin around the base of the tail. Should the animal's tail be pinned by a predator, the skin breaks, allowing the panicked rat or mouse to literally slip its skin, pulling vertebrae, muscles and tendons out and leaving its limp tail skin behind. Unfortunately, this adaptation does not save the tail, which soon dries up and drops (or is chewed) off; unlike many lizards and amphibians, rodents cannot grow a new one. But the individual survives to reproduce another day.

In general, rodents use escape strategies rather than defensive ones. The most notable exception is the porcupine. If cornered or caught, most rodents will attempt to bite with their formidable incisors. I once watched a woodchuck fend off my 100-pound Alsatian, which had pursued it to the center of a soccer field. The woodchuck was holding its ground, growling and loudly clacking its incisors as it stared down the dog. The dog kept glancing back towards me, clearly in need of instructions. (I think she hoped I would call her away from the embarrassing standoff.) During one of these turns the woodchuck launched itself into the air and connected with my dog's nose, then hung from her lip for a few seconds before dropping to the ground to make an unchallenged run for the bushes. The incident was the start of a lifelong grudge match; our property is now free of woodchucks.

Although most rodent species tend to specialize, there are a few successful generalists. Most notable are the introduced murids, the Norway rat, black rat, and house mouse, all of which have been more successful than many native species and, in some cases, have actually displaced them. They are unusual generalists in that they seem to succeed at nearly everything they do. The silver rice rat, endemic to the Florida Keys, has specialized itself for mangrove and salt marsh habitats. This niche, however, has been overtaken almost entirely by the black rat, which occupies the same regions and eats the same foods. The critical difference between these species is that the black rat also proliferates in upland thickets, buildings, and dumps. Storm flooding and hurricanes periodically decimate shoreline habitats in the Keys, wiping out coastal rat populations both native and introduced. Unlike rice rats, however, black rats have a reserve population. Their relatives in sheltered upland sites are able to survive the storms; they later reinvade shore areas when the coast, literally, is clear. In the 20 years that I have studied Florida Keys rice rats, their populations have declined relative to those of the black rat—a striking failure for a species in its own habitat.

A combination of genetic flexibility and brain power may account for the invader's success. One characteristic, "trapability," illustrates a basic difference between the two species: I have always found it difficult to catch a black rat twice, but can easily capture rice rats over and over again. I gained some insight into this difference when I trapped a black rat and a silver rice rat in the same mangrove swamp. I glued a small capsule containing green and pink fluorescent dust to the neck of each animal. The capsules are designed to leave a dust trail as the animals move, which I can follow with an ultraviolet flashlight. Though it doesn't hurt them, the tagging process is an experience that always panics both species.

I released both captives after dusk where they had been trapped. Several hours later, I began to track their trails, crawling through, around, and over the prop roots of mangroves. Illuminated by the purplish light of my ultraviolet flashlight, the lines of fluorescent powder left by the rats glowed along branches and spidery prop roots, tracing silhouettes of the forest's framework. The trails revealed to me that each rat soon had encountered other traps that I had set. The black rat's dust trail headed straight toward one, but stopped about six inches short of the entrance. Trampled dust at that point showed that the animal had paused. Its trail then abruptly departed from its earlier course, describing a perfect circle around the trap. At several spots I saw small clumps of fluorescent dust; apparently the rat had paused each time to eat a few stray flakes of oats that I had dropped while setting the trap; then it veered off deep into the mangroves. It was clear to me that the black rat had been tempted by the bait but, like so many of his kind, had learned from the unpleasant experience he'd had earlier that day. Not so with the rice rat; it was hunkered forlornly in another of my traps, the bait long gone. A beeline of dust showed that it had headed directly for (and through) the trap's open door, throwing all caution to the winds.

The relative naïveté of that rice rat is an apt example of the vulnerability of all specialized native species. Silver rice rats join a long list of animals worldwide whose populations are being adversely affected by species that humans have introduced. Unless we can control black rats more effectively than we have in the past, the silver rice rat is likely to become extinct, just as the endemic Galápagos and Antillean rice rats succumbed, following the introduction of black rats to those islands. Once a species disappears, it can never be recreated; if we wish to preserve the marvelous diversity of North America's fascinating rodents, we must learn to manage and protect them. ■

Family Aplodontidae Mountain Beaver

Only one living species of this primitive family exists: *Aplodontia rufa*, mountain beaver. Neither a beaver nor restricted to mountains, it weighs one to three pounds and is highly fossorial, with long and heavily clawed forefeet for digging. It has small, gopherlike eyes and ears, and long facial whiskers that serve much the same function as a blind person's cane, helping it navigate the dark underground tunnels that it digs and maintains. Despite extensive adaptations to a subterranean life, it also will climb shrubs and trees—though somewhat awkwardly—to harvest twigs. It occurs in the Sierra Nevada and along the Pacific Northwest coast.

Family Sciuridae Squirrels

As a family, squirrels form quite a diverse group, including species that range from burrowing (prairie dogs, ground squirrels, chipmunks), to arboreal (red and gray squirrels), and even to gliding species (flying squirrels). Most squirrels are active primarily during daylight hours, although flying squirrels are nocturnal.

Vision is an important protective sense for all members of this family, and even the burrowing species have large eyes. "Sciuridae" means "shadow-tail," and apart from the more fossorial species, the tail tends to be long and often bushy. Arboreal and gliding species use it for steering, as a counterbalance, or as an airfoil. Flying squirrels are the most highly modified, also possessing a loose skin flap on each side of the body, between the fore and hind legs. They do not flap these "wings" as a bird does, but rather leap into the air and use them to glide in kitelike fashion from tree to tree. Some species, such as marmots, hibernate for many months at a time; others, such as gray squirrels, do not hibernate at all. Many species are notorious hoarders of seeds, grasses, nuts, and fungi; some also eat insects.

Gray Squirrel *(Sciuris carolinensis)* This handsome gray squirrel flashes his paired upper and lower incisors, a distinguishing characteristic shared by all rodents. It is one species that has managed to thrive alongside humans. New England's virginal forests of two or three centuries ago, for example, were dominated by beech trees—at best, an ephemeral source of food for gray squirrels. But then Europeans cleared those forests for timber and farmland. Generations later, as farming declined there, many cleared areas began to grow back. Today, the beech-dominated climax stage has yet to return, and recovering forests consist largely of oak and hickory trees, important food resources for gray squirrels.

When nuts and seeds are plentiful, gray squirrels hoard them, often burying what they would otherwise eat. If they forget their hiding places or have so much food that digging up buried items becomes a waste of time, those seeds can germinate, creating future food plants for them. In this way, squirrels influence forest composition and play an important role in their environment. Forest or no forest, this animal has survived throughout the eastern half of the U.S. into Canada; in developed suburban areas, bird feeders have given it a boost.

Gray squirrel shows off a trademark of the order: ever growing incisors

Eastern chipmunk packs its pouches

Abert's squirrel gleans the ground

Eastern Chipmunk (*Tamias striatus*) Chipmunks dwell alone in burrow systems that can run underground for tens of feet. Why the gaudy stripes? Like many other ground squirrels, chipmunks have developed horizontal pin-striping, a common form of camouflage that confuses predators by tricking their eyes into falsely extending the lines and misjudging the squirrel's position.

Since they are ground squirrels, chipmunks are extremely vulnerable when venturing away from their burrows to forage. One solution to this dilemma has been to develop fur-lined cheek pouches that hold half a dozen acorns at once. On highspeed foraging missions, chipmunks efficiently tamp desirable nuts and seeds into their cheeks until the cavities are stretched to capacity, at times causing their heads to appear twice normal size. They may then bolt, squeaking and scolding, toward the safety of their burrows, where they can eat in peace.

Abert's Squirrel (*Sciurus aberti*) Tassel-eared and fluffy-tailed, this arboreal squirrel occurs in the southwestern U.S. and Mexico, inhabiting forests of ponderosa pine, which serves both as food and nesting habitat. Accuracy is critical when leaping between wind-tossed treetops, and a luxuriant tail helps the Abert's steer itself in mid-air. Gauging distances precisely is also vital, but squirrels cannot use binocular vision in the same way that many other mammals do. Primates, cats, and bats, for instance, have eyes closely spaced on the frontal plane of the face; because the focal fields of both eyes overlap, their vision is stereoscopic, or three-dimensional. Squirrel's eyes, positioned on the sides of the head, help them sense predators from all directions but offer little overlap. Even so, they have learned to judge depth by using the relative movement of objects around them. Before a leap, the squirrel bobs its head, making nearby objects appear to shift more than distant ones and giving it information on relative distances within its field of view.

Red squirrel

Harris' antelope ground squirrel

Red Squirrel (*Tamiasciurus hudsonicus*) Harvesting spruce cones when green and then waiting for them to ripen is this squirrel's specialty, although it eats almost anything, from young birds to shed antlers to maple sap. Ranging from the Carolinas and New Mexico to Alaska, it overlaps the distribution of the gray squirrel in places, but the two are not competitors. For while the gray relies largely on acorns and hickory nuts, the red prefers the seeds of conifers. This small but important distinction allows the two to closely coexist without conflict.

Harris' Antelope Ground Squirrel (*Ammospermophilus harrisii*) Perhaps the most conspicuous and entertaining mammals in the dry areas they inhabit, antelope ground squirrels characteristically scamper and scurry about nervously, tails upright or arched over the back. They are burrowers.

Harris' antelope ground squirrel, like other species of this group, has a short and scrawny tail. A luxuriant one would only get in its way, for this is no arboreal acrobat. Instead, it spends most of its time rushing about the ground, foraging on arid and sparsely vegetated plains and slopes. Like other squirrels, it uses its tail as part of an alarm display, holding it straight up and chirping loudly as it runs. Living in areas without much protective ground cover, this species is highly visible to predators, especially since it is diurnal. To counteract the risk of predation, it excavates and maintains several escape burrows within its territory; this squirrel is never more than a short dash from safety.

Like chipmunks, Harris' antelope ground squirrels sport head-to-tail striping as camouflage and are equipped with cheek pouches. They store what they find in burrows, under rocks, and in other shelters. They are active throughout the year except in cooler areas of their range, where they become inactive in winter but do not actually hibernate.

Family Geomyidae Pocket Gophers

Geomyidae—"earth mice"—include dozens of species and subspecies commonly called pocket gophers, for their furry cheek pouches. They are poor dispersers; a river or even a long stretch of rocky soil can isolate different groups from each other. Therefore, many hundreds of distinctive pocket gopher populations have developed, each with its own coloration and other features.

Northern Pocket Gopher (*Thomomys talpoides*) Like other members of its family, this species commonly forages underground within a spidery system of tunnels. Equipped with tiny eyes and ears, sleek fur, and sturdy front legs with long, curved claws, pocket gophers are better adapted to a subterranean existence than any other North American rodent. Underground, they can scoot backward almost as fast as they do forward, using the highly sensitive tip of their tail to feel the way. Their teeth serve as an additional digging tool; they avoid mouthfuls of dirt by folding their furry lips behind the gnawing incisors. Foods include bulbs and roots browsed from a network of shallow tunnels they dig. Observers at the surface have reported seeing an entire plant apparently sucked into the ground as a pocket gopher pulls it, leaves and all, into its runway below.

Banner-tailed kangaroo rat

Family Heteromyidae **Heteromyids**

Banner-Tailed Kangaroo Rat *(Dipodomys spectabilis)* Here is a species that has become so well adapted to desert living that, in spite of its arid environment, it does not need to drink any liquid water. Banner-tailed kangaroo rats get all the moisture they require from seeds and grains collected on the desert floor, since they are able to subsist on the relatively tiny amount of metabolic water that is produced as they break down carbohydrates. They also minimize their water needs by restricting activity to nighttime, when it is cooler. During days and extremely hot periods they estivate, lowering their metabolic rate—and therefore their water consumption—by entering a state of torpor. A subterranean existence also helps retard water loss, since the burrows are more humid than open air.

Kangaroo rats typically maximize that humidity by sleeping with their noses tucked into their fur, which increases condensation. In addition, they have highly specialized and layered nasal membranes that are the anatomical equivalent of cooling fins. As the animal exhales, water vapor in its moist, outgoing breath condenses on the cool nasal membranes and is resorbed into the body, minimizing water loss. Lastly, kangaroo rats possess extremely efficient kidneys, which produce a highly concentrated urine, further minimizing any water loss.

Beaver slaps its tail to warn of danger

Family Castoridae Beaver

Beaver (*Castor canadensis*) Beavers can eat every part of a tree but its woody core. Before sitting down to lunch, however, they must find and fell a suitable tree or sapling, gnaw it into manageable sections with their large, chisel-like incisors, perhaps drag the pieces to a canal they have previously dug, and then raft them downstream to their home pond. Beaver families are nature's equivalent of the Army Corps of Engineers—they seem obsessed with modifying their environment and are notorious builders of dams, which they use to block streams and create large ponds around which they live and forage. A beaver pond must be sufficiently deep to allow the animals to construct their lodge, a large mound of sticks and mud often surrounded by water and equipped with underwater entrances that lead to interior living chambers, above waterline. These aquatic rodents spend half their

Beaver slices apart an aspen

time swimming and much of their energy altering their surroundings; the lodge is their base of operations. They sleep and overwinter in it, storing food nearby. They are such avid dam builders that they will sometimes construct an elaborate dam in a pond with no inlet or outlet, creating a structure that serves no purpose other than to subdivide the pond.

Beavers have webbed feet hind and fore, and the tail is a horizontally flattened pad that is so scaly that 18th-century clerics approved the eating of beaver meat on fast days. The tail is used for swimming, both as a rudder and sculling oar, and for signaling alarm (by slapping the water's surface). Beavers can remain underwater as long as 15 minutes, because their heart rate slows as they dive, lowering their bodily needs for oxygen. They typically weigh 40 to 60 pounds and measure 3 or 4 feet in length, but one giant trapped in Wisconsin tipped the scales at 110 pounds.

Family Muridae Rats and Mice

Worldwide, some 1,300 species of rats, mice, hamsters, gerbils, voles, and others belong to this, the largest rodent family—indeed, the largest family of all mammals. Native to every continent except Antarctica, murids eat mostly seeds and vegetable matter, although some forage for insects, eggs, or even fish or carrion. All have 16 teeth. Those of Canada and the United States number about 100 species, and fall into two groups: pointed-nosed rats and mice, and blunt-faced voles and lemmings.

White-Footed Mouse (*Peromyscus leucopus*) Learning by mistakes, a white-footed mouse nibbles a monarch butterfly (above). It soon will avoid such prey, associating the butterfly's warning coloration with its bad taste. This murid and many others exhibit cryptic counter-shading: Seen from above, their dark backs blend with the ground; when climbing and seen from below, their creamy white stomachs disappear against the sky. Long whiskers and sensitive ears help them detect the numerous raptors, snakes, weasels, foxes, cats, and others that prey on them. Parental care is minimal; a mother will rapidly abandon her nest if she senses an approaching snake. Her blind and furless young, however, clamp so firmly to her teats that they are dragged with her to safety. This mouse is successful throughout much of its huge geographic range, which includes the eastern and midwestern U.S. and eastern Mexico, south to the Yucatan Peninsula.

M u s k r a t *(Ondatra zibethicus)* This scratching muskrat may be covered with fleas. For although it spends much of its time in water, its dense, waterproof coat traps air, which insulates its body and can harbor parasites. Largest member of the murid family, the muskrat is named for its ratlike tail and odoriferous musk glands. It occurs in freshwater marshes, ponds, and streams from Alaska to the Gulf of Mexico, and, like the beaver, builds lodges and is prized by trappers. Its feet, however, are not webbed, and its tail is somewhat flattened vertically. Muskrats are not closely related to beavers; think of them as large, amphibious field mice that weigh about three pounds and measure up to two feet long.

Family Zapodidae Jumping Mice

North America's four species of jumping mice have adopted a two-legged approach to success. Like kangaroo rats, kangaroos, and some other mammals, members of this group have large hind feet and legs, long tails for balance, and they rely on the fast, unpredictable movements that result from hopping. Their erratic, ricocheting motion makes them a tough target for many of the usual rodent predators—owls, snakes, weasels, foxes, and such—and they can leap six feet in a single bound. Their tricolor coats—brown back, yellow-to-orange sides, white belly—provide them with fine camouflage in the meadows and woodlands they inhabit.

New World Porcupines

Porcupine *(Erethizon dorsatum)*
Unfazed by the thorns of a wild rose, these porcupines revel in the greenery of spring, gorging on herbaceous vegetation, tender twigs, and tree buds. Throughout winter, they exist on tree bark and evergreen needles. Special bacteria in their intestine help them absorb nutrients from otherwise indigestible cellulose.

Beneath each animal's soft guard hairs lurks an arsenal of stiff, five-inch-long quills. A porcupine will erect and rattle its quills when challenged, and often will turn its backside toward its foe, duck its unprotected face, and lash its spiny tail back and forth. The hollow, loosely attached quills are easily dislodged, and replacements soon grow in. Minute barbs on these modified hairs anchor them into the skin of the would-be predator, whose subsequent movements often ratchet the quill in more deeply. Quills can be fatal if they puncture a vital organ.

While porcupines tend to be solitary and even somewhat territorial much of the year, during winter they may share dens or nests in caves, hollowed logs, or trees. They may choose to avoid bad weather but do not hibernate. They are principally nocturnal but can be active in the day even under the worst conditions.

Of all North American rodents, only beavers are bigger; porcupines average three feet long—including a six-inch-long tail—and about 15 pounds. They range from Alaska and northern Canada to the southwestern U.S. and Mexico.

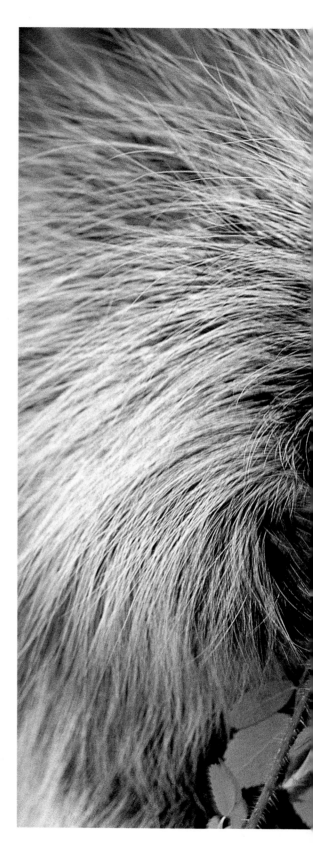

Porcupine and offspring nibble a wild rose

Cetaceans

Cetaceans

Order Cetacea

BY KENNETH S. NORRIS

Sixty-five million years ago, a five-and-a-half-mile-long asteroid powered into the Yucatan Peninsula of Mexico, throwing up huge amounts of dust and debris that blotted out the sun with an opaque pall that swept around the world. The reigning dinosaurs of that era died, as did many other forms of life. The mammals that managed to survive this holocaust were mostly small and nocturnal. When the skies cleared, they found themselves in a nearly vacated world full of opportunity.

They proliferated into several lineages, including a group of little antelope-like creatures now considered to be the ancestors of modern whales and dolphins.

Some of these animals, called mesonychid protoungulates, became amphibious and developed stabbing canine teeth—perhaps to capture fish in the estuaries of the warm, nearly landlocked Tethys Sea of that time. It is in the ancient sediments of the Tethys that the earliest fossils of cetaceans have been found.

Leaving the land for the sea was no easy transition. Nothing much worked well for those first mesonychids. Their balance was all wrong and their legs were poorly designed for swimming. When they first ducked their heads under water, their vision was a blur; they could not locate things by sound. But over millions of years, their legs shortened and began to disappear; they began to swim with an up-and-down movement of trunk and tail. Lungs shifted up under the spine, improving their undersea balance. Eyes became modified to see well through water. Ears, too, became adjusted to undersea life, a crucial development because hearing is even more important underwater than it is in air. In the sea, good hearing enables an animal to quickly scan beyond the limited reach of sight.

The mesonychids gave rise to the most ancient group of true cetaceans, the archaeocetes, which spread throughout the Tethys. One branch gave rise to the almost snakelike Basilosaurus, whose fossils occur widely in Tethyan deposits, including those of North America. Another group of archaeocetes, the streamlined dorudontids, looked superficially much like modern dolphins. Their tapered, bullet shape conserved body heat and suited these animals for cool seas. Indeed, the dorudontids seem to have escaped the warm Tethys about 40 million years ago, quickly spreading almost from pole to pole. With that escape, the direct ancestors of modern cetaceans appeared. Two major lineages began to take form.

One would lead to modern baleen whales, what we now call the suborder Mysticeti, often huge animals that take whole schools of prey into their mouths at a time and then force the seawater out through fibrous, sieve-like baleen that hang down from the upper jaw. The other lineage, the suborder Odontoceti or toothed whales, differentiated into the many dolphins, porpoises, and smaller whales. A new sense—not just hearing but echolocation—also appeared.

Modern odontocetes echolocate by directing out into the sea collections of clicks, some very high pitched, far beyond the upper level of human hearing. The clicking animal then interprets the faint echoes reflected back from any object in the water. Even ancient toothed whales possessed an inner ear apparatus capable of high-frequency hearing. Both the first mysticetes and odontocetes had teeth; only

previous pages: **Killer whale and dolphin cruise side by side** over much time did the baleen whales lose them. Even today, some baleen whales temporarily possess primordial teeth during certain fetal stages.

Baleen whales are generally much larger than toothed whales. The exception is the sperm whale, a huge creature of the open sea with rows of teeth studding its lower jaw. Sperm whales eat primarily deep-sea squid and octopuses; to capture such prey, they make prodigious dives to depths never touched by sunlight.

Mysticete whales eventually split into three major lineages:

(1) Skimming whales, represented today by right and bowhead whales, swim with mouths partially open, taking in sea water and plankton as they let the water flow out through their very long baleen.

(2) Gulping whales, such as fin, blue, sei, and humpback whales, seek schools of plankton, squid, or fish. They gulp huge mouthfuls into distensible throat pouches, then force the water out through their baleen with their tongue.

(3) A single living species, the gray whale, is termed a grubbing whale. Typically, it feeds by plowing its snout several inches into a mud bottom and ejecting a mouthful of water into the pit it has created, liquefying the sediments. The whale then sucks in the resultant slurry of mud, bottom-dwelling shrimps, worms, amphipods, and other invertebrates, and forces this mix upward against its baleen, allowing the watery mud to exit while the prey is retained and swallowed.

Modern cetaceans include about 80 species, 47 of which swim in North American waters. Each exploits a particular food source. Most baleen whales migrate between warm-water calving grounds and abundant cool- or cold-water food sources in the high latitudes. Toothed whales inhabit all oceans of the world, and some even live in freshwater river systems. They use their keen echolocation sense to find and track prey, both individuals and schools.

North American cetaceans range in size from the tiny Gulf of California harbor porpoise, only about 5 feet long, to the gigantic blue whale, which tops out at about 100 feet in length. There are no cetaceans as small as squirrels or mice; the power of water to suck away body heat precludes such diminutive body size.

Our core knowledge of cetaceans has been achieved with great difficulty, for these creatures are of the sea, while we are of the air and land. Also, our perceptions of whales and dolphins have been colored by the men who hunted them and by artists who, of necessity, based their works on the reports of seamen.

Slowly, humankind realized that whales and dolphins could be remarkably accepting of non-threatening humans in their midst, and both scientific and popular literature began to reveal a new, more benign image of these creatures. It is only relatively recently that scientists, photographers, and artists have been able to produce truly lifelike representations of cetaceans. Photographers have been the most intrepid ones, venturing into arctic waters as well as the open sea.

Knowledge of ceteacean anatomy has proved easier. If a scientist could stand the stench of rotting whale, he could have all the material he needed just from stranded carcasses on the beaches. Early on, dissection and observation led to some magnificent folio works describing the skulls, bones, baleen, and even the soft anatomy of whales, such as their vital organs and blood vessels.

More recent research involving live whales has taken two basic approaches: Captive animals kept ashore in aquariums where they can be closely observed, and wild animals at sea, observed by a new breed of naturalists who roam the oceans with their subjects. Captive studies have yielded much information about cetacean behavior and physiology, how cetacean senses work, and various qualities of the cetacean mind. Research in the open sea has taught us where cetaceans live, how they hunt, what their populations are like, how they reproduce, and more.

One early milestone in cetacean research involved a captive school of bottlenose dolphins held in a public exhibit—Marineland of Florida, at St. Augustine. To the surprise of many scientists, these animals exhibited a social structure very similar to that of many terrestrial mammals; they courted and bred seasonally, and their pregnancy and nurturing proved long, not terribly different from our own. Fully active young were born only after a long gestation, and were tended with solicitous maternal care through sometimes rambunctious childhoods.

Behavioral studies of wild cetacean societies came later. Two such efforts began in the late 1960s, one with a nearshore bottlenose dolphin population off Sarasota, Florida, which I was fortunate to head. The other concentrated on oceanic spinner dolphins near Hawaii. Both projects owed much to Jane Goodall—that patient, unobtrusive, pioneering naturalist who set up camp in an African jungle and lived among a troop of chimpanzees until she knew every one. By showing that such studies were possible, Goodall's work encouraged not only other primatologists but all sorts of biologists, including myself, to pursue similar observational efforts.

Dolphins live their entire lives inside fluid schools that provide protection and social structure. They are born into these schools, are nurtured there, become adults, and finally die—all within the school's confines. Most dolphins never stop moving, even when they rest. They rely not on ferocity for defense, rather on the subtle protection provided by schooling and by their echolocation sense, which gives them early warning of approaching predators.

The Sarasota study of bottlenose dolphins yielded valuable information on dolphin requirements. It also convinced scientists that these animals needed protection from human thoughtlessness, including boat traffic over dolphin home grounds and onshore construction that destroyed traditional food sources.

Both spinners and bottlenose dolphins have well-defined visual limitations. Bottlenose dolphins can see only a few feet through murky bay waters; the night-time world of spinner dolphins is as black as a pocket. So for both species, sound has become the prime sense. Young dolphins learn to echolocate by mimicking older members of the school. Each dolphin's call has a certain individuality, and seems to bear the marks of relationships with other dolphins. Each animal knows its home grounds in much detail, and will find its way back even when it has been displaced miles away.

Studies with baleen whales also have told us much. We now know that male whales, especially humpbacks and bowheads, sing songs—stanzas of groans, moans, and chirps, repeated almost endlessly—while on their mating grounds. These calls seem to play a role in how the males space themselves out and attract mates. Why the enormous complexity of their songs? No one knows.

Using U.S. Navy underwater listening arrays that once guarded America's shores from Soviet submarines, one cetologist has recorded the sounds of the great baleen whales from more than 100 miles away. Presumably, whales also hear each other at such distances. If so, the entire ocean becomes for them a vast listening bowl—and their ability to migrate great distances becomes less puzzling for us.

Captive bottlenose dolphins have become such a part of our world that some can be considered domesticated. Studies indicate that the bottlenose dolphin mind is in many respects a good one. Dolphins can learn to respond to queries posed by short strings of symbols, arranged much like words in a language. In fact, they can remember lists of symbols as well or better than we do. One researcher, using standard animal training techniques, asked a dolphin to respond to the abstract request, "do something new." Eventually it poured out behavior patterns no one had ever reported before—graceful back dives, spiral swimming, and much more.

Experimenters have found dolphin echolocation to be the most refined such sense in the animal kingdom. Trained dolphins, using long series of remarkably intense clicks, have located submerged targets at unbelievable distances—objects the size of tangerines sensed more than a football field away. Also, they can discern differences between sheets of different metals—copper, aluminum, and steel—even when painted the same color. One trained dolphin dove nearly 1,000 feet down to press a signal lever. Photos show the animal's highly flexible chest cavity collapsing under the immense pressures, only to expand to normal shape as it surfaced.

Dolphins have been found to sleep one eye at a time, switching sides halfway through sleep periods. They sleep in groups, always orienting themselves so that the open eye is toward their companions. Such behavior certainly makes it seem that they are keeping track of each other, while more alert members continuously echolocate into the surrounding sea, seeking signs of predators.

A new breed of cetologists, taken together with an evolving set of new approaches, has enabled the science of modern cetology to begin to tell—at times in remarkable detail—what has become of those early protoungulates that began to dabble in the estuaries of the Tethys Sea so long ago. ■

Family Ziphiidae Beaked Whales

Named for their tapering, dolphinlike snouts that, in some species, resemble the neck of a bottle, beaked whales are medium in size, have small dorsal fins and flippers, and usually possess only one or two pairs of teeth. They hunt squid and deep-sea fish. Ten species of beaked whales frequent North American waters.

Family Physeteridae Sperm Whales

Three living species compose this family: pygmy and dwarf sperm whales, which are not known to exceed 11 and 9 feet in length, respectively, and the true sperm whale, males of which grow as long as 66 feet. All three frequent North America's major coasts. Massive foreheads and narrow lower jaws mark this family, giving members a "chinless" appearance. The blowhole is located well to the left of the midline, and the spout is directed at about 45 degrees to the horizontal. All are prodigious divers, wandering far down into the dark abyssal sea, where the only light that illuminates predator and prey is biological in origin. They feed primarily on deep-water squid, octopus, and fish. Sonar studies have tracked true sperm whales swimming as deep as 8,500 feet, more than a mile and a half straight down.

Moby Dick in the making? A mother sperm whale escorts her rare albino baby

Sperm Whale *(Physeter catodon)* The lineage of this large diving whale reaches back almost to the split between the earliest baleen and toothed whales, when all cetaceans had teeth. Its enormous head accounts for as much as 40 percent of its body size. Within lies a reservoir of transparent, liquid wax—spermaceti—that the animal may use to propagate sounds. This was the great prize of the sperm whaler; it could be used to make candles that burned with an especially bright, if smoky, flame. It also served as a peerless lubricant for the machines of man.

Sperm whales suffered from excessive hunting, but remain the most numerous great whale, numbering perhaps two million worldwide. Only males wander to the extreme latitudes north and south, leaving the females and young in more temperate waters. They are polygamous, grouping themselves roughly by age and sex.

Family Monodontidae Monodontids

Beluga *(Delphinapterus leucas)* Truly a creature of the ice, the white-skinned beluga has been radio tracked to nearly 75° north, where its only access to air is through long "leads," or cracks, in pack ice. Its head is highly mobile, thanks to neck vertebrae that are not fused, as they are in many whales. Like the narwhal, it lacks a dorsal fin—an advantage for an animal that swims long distances under polar ice. It ranges as far south as Alaska's Bristol Bay and the Gulf of St. Lawrence.

Belugas can grow up to 16 feet in length and attain perhaps 3,500 pounds in weight, with males somewhat larger than females. They feed on salmon, capelin, pike, and other fish, as well as invertebrates. In midsummer, they enter rivers, prompting the outer skin to swell and slough away during the yearly molt. At this time they are especially vulnerable to attacks by their principal predator, the polar bear, which can kill a beluga with a single well-directed bite. They have been seen well up the Yukon River, as far as 600 miles from the sea. In captivity, the white whale has proved itself both docile and endearing.

Squadron of narwhals

Narwhal *(Monodon monoceros)* The narwhal's single, magnificent tusk of twisted ivory—which can grow up to nine feet long—gave rise to the unicorn of European myth. In time the tusk itself took on mythic properties and was included in potions to ward off epilepsy. Since inbreeding among Europe's royal houses fostered a high occurrence of epilepsy, the tusks were especially prized by royalty.

As exploration of the Arctic proceeded, part of the mystery was soon cleared away—the narwhal's tusk is one of only two teeth that adult male narwhals have, protruding through the upper lip like a great twisted taper. Only recently have humans ventured close enough to narwhals to observe how males use this tooth as a great spear in sometimes fatal combat, as rivals joust or try to drive it with great force into an opponent. Also creatures of the ice, narwhals have been spotted as far north as 85° N. About the same size as belugas—exclusive of their tusk—they have fan-shaped flukes and a mottled pattern that makes them all but invisible as they swim beneath broken ice. Males, especially, grow lighter with time; older ones may become almost entirely white and are considerably larger than females.

Family Delphinidae Delphinids

Most abundant and varied of all cetaceans, the world's delphinids include about 40 species of dolphins, porpoises, and small "whales." The group is defined mainly by skeletal characteristics; its member species are oceanic, feeding primarily on fish in upper waters. They are toothed, and most have a dorsal fin and a notched fluke.

Bottlenose Dolphin *(Tursiops truncatus)* Perhaps the world's best known wild animal, the bottlenose is familiar to millions who have seen it perform in oceanariums. Possessing a brain nearly as large as ours relative to body size, this cooperative creature has been trained to answer questions such as, How deep can you dive? (at least 1,000 feet), How fast can you swim (at least 21 miles per hour), and How well can you remember strings of numbers (about one digit better than we can). A generalist, the bottlenose inhabits warm ocean waters and bays, where it grows to 10 or 12 feet and usually weighs in at under half a ton.

previous spread: **Bottlenose dolphins cavort off Hawaii**

Heavily scarred Risso's dolphin breaks the surface

Dall's Porpoise *(Phocoenoides dalli)* This swift-swimming cetacean with striking, black-and-white coloration ranges from the Bering Sea to Baja California. It travels in groups of up to 20 animals, often racing alongside the bow of a moving ship, seemingly daring the pilot to run it down. Somewhat smaller than the bottlenose, it has suffered from the Japanese commercial fishery, which has killed as many as 60,000 Dall's porpoises yearly, 40,000 deliberately and 20,000 by accidental entanglement in their North Pacific drift net operations.

Risso's Dolphin *(Grampus griseus)* Big and boisterous, this dolphin occurs in all warm and temperate waters of the world. One individual, dubbed Pelorus Jack, escorted vessels into a New Zealand harbor for some 20 years.

Adult Risso's dolphins can reach 12 feet long and perhaps 900 pounds. Most are so scrawled with prominent scars—believed to be caused by the teeth of their own kind, or possibly by squid—that individuals can be easily identified. Older animals acquire so much scar tissue that parts of their bodies appear nearly white. They are thought to feed primarily on squid and often hunt deep coastal waters, especially areas with rugged undersea topography, where such prey gather. They lack a defined beak but possess a large hooked dorsal fin and a deep groove running down the middle of the forehead. While the furrow's function is unknown, educated guesses associate it with some aspect of echolocation.

Killer Whale (*Orcinus orca*) Largest member of the dolphin family, the killer whale or orca runs as big as 30 feet and 10 tons, with mature bulls topping the charts. True to its name—and unlike most other cetaceans—it preys on other warm-blooded animals; it is the marine counterpart of the tiger or wolf. Like them, it is both agile and fast, attaining speeds of nearly 30 miles per hour.

Killer whales live in very stable, long-lived family units called pods, often centered around an old female. Like wolves, they move in regular seasonal patterns; unlike land predators, however, they prowl continuously, day and night. Each pod seems to specialize in a particular food source, some hunting whales or seals, others preferring salmon. Members of one pod use a dialect easily distinguishable from

Young and old orcas patrol a Pacific shore

Orca pursues a Dall's porpoise off Alaska

those of other pods; groups that hunt seals are almost silent, while some salmon specialists are extremely talkative. Killer whales can mount coordinated attacks that enable a pod to take large prey, such as a young gray whale, with some members biting or slashing the whale's flukes while others try to cover its blowhole.

As with other top predators, the total number of killer whales in a given area may be remarkably low; some groups migrate over long distances, preying on young marine mammals that are born in specific and often disparate locations. Worldwide in distribution, killer whales range from the edges of one polar ice cap to the other, especially near coastlines. They may live 70 years or more, all the while keeping to the same family groups.

Family Eschrichtiidae **Gray Whale**

Gray Whale *(Eschrichtius robustus)* Sole living member of the family Eschrichtiidae, the gray whale is considered a "living fossil"; its skull is designed along the lines of whales that lived millions of years ago. Today it roams the North Pacific from the Chukchi Sea to Baja California, where it breeds and calves. Once, perhaps as recently as the 18th century, it also ranged the North Atlantic.

Gray whales grow up to about 50 feet long and 35 tons. Undoubtedly North America's most watched whales, they have been seen by millions along the populous and easily accessible West Coast, where they frequent shallows, often feeding inshore among the rocks and kelp. Their "grubbing" mode of feeding—ramming into mud and ejecting water to make a slurry from which they filter out

Gray whale surfaces off the Pacific coast

invertebrate prey—is unique; as they feed, they leave the bottom stitched with lines of depressions, the water clouded with mud.

Gray whales mate in late November and early December. Typically, pregnant female grays spend the summer months in Alaskan waters, heading south in September to reach their calving grounds by December, and give birth soon after. They are extremely protective and defend their young ferociously. Not all of them follow the established migration pattern, however, and gray whales can be seen throughout the year in the Gulf of California, off the Farallon Islands near San Francisco, off Vancouver Island, and in other areas. In 1972 Mexico created the world's first national whale refuge, setting aside Baja California's Laguna Ojo de Liebre (also called Scammon Lagoon)—to protect the gray whale nursery there.

Blue whale mother and calf

Family Balaenopteridae **Rorqual Whales**

The term "rorqual," of Norwegian origin, refers to a family characteristic: pleated, expandable throats that enable members to gulp whole schools of prey in a single mouthful. They include six of the world's biggest and swiftest species, all of which roam North American waters: the blue, fin, sei, minke, Bryde's, and humpback.

Blue Whale *(Balaenoptera musculus)* This largest of all mammals (about 200 tons) is also the slimmest great whale, thanks to its great length (up to 100 feet). It feeds on shrimplike krill and can down over a ton at one feeding. Early whalers could not capture blues due to their great speed and their propensity to sink once harpooned. But the invention of harpoon guns and fragmentation bombs enabled whalers in the age of steam to pursue blues nearly to extinction. The killing ended in 1966; only now are some blue whale populations coming back.

Humpback breaches near southeastern Alaska

Humpback Whale (*Megaptera novaeangliae*) Best known for its haunting songs and photogenic breachings, the amiable humpback grows to about 50 feet in length and 50 tons in weight. Once one of the world's more abundant great whales, it was decimated by whalers until 1966, when protection was granted to the few thousand left. Like other large whales, humpbacks migrate throughout their lives, alternating between high-latitude feeding grounds and warmer breeding and calving areas, where males space themselves out, apparently by song, to pursue and couple with receptive females.

Humpbacks eat mainly krill but also small fish, corraling the latter in bubble nets: The whale streams out a band of bubbles as it circles a school of prey fish. The fish, unwilling to pass through this bubble curtain, remain tightly concentrated, enabling the whale to dive and come up inside the curtain, mouth agape—and feed on the whole school. At times, several humpbacks will fish together.

Family Balaenidae Right and Bowhead Whales

Termed right whales by whalers because they were slow moving, rich in oil, and didn't sink after being killed, members of this cetacean family are chunky in shape, with a huge skull that makes up to 40 percent of its length. Like the rorquals, they are filter feeders. But they lack the rorquals' throat pleats, relying instead on highly arched upper jaws and scooplike lower jaws that give their mouths great capacity. Indeed, the bowhead's name reflects the extreme arch of its jaws. Its huge, powerful head helps it break through Arctic pack ice for air. Whalers judged bowheads the most valuable species, for they provided immense amounts of oil as well as the finest baleen, or "whalebone," used as corset stays, umbrella ribs, and fishing rods.

Flukes emerge just before a right whale dives in the Bay of Fundy

Right Whale *(Balaena glacialis)* Once numbering in the hundreds of thousands, right whales were nearly exterminated before being protected in 1935. Today only a few hundred remain in the Northern Hemisphere, about 1,500 range the Southern Hemisphere. It was in a southern population that whale song—a seamlessly repeated series of sounds, arranged much like a piece of music—was first discovered. Right whales—principally males—make a remarkable variety of deep moans, whoops, and higher pitched sounds, repeating their entire "score" over and over when they enter the breeding grounds. The songs can be heard for hundreds of miles, and seem to help the animals organize themselves. Similar songs have also been recorded for humpbacks and bowheads.

Carnivores

Order Carnivora

BY JOHN SEIDENSTICKER

I prudently approached bottom in a steep-sided Idaho canyon. Readings taken from the ridge above told me that the radio-collared puma I had been tracking, an adult without kittens, had lingered here for two days. I suspected she might be with a kill. Her radio signal now seemed to come from a mile or so down canyon. Then I saw, under a fir tree, the remains of a bull elk that the cat had carefully covered with fir needles and sticks. Deep croaks of ravens and the sharper calls of magpies and jays filled the tight canyon as I moved up to examine the carcass.

A dead, partially eaten elk covered with sticks and needles is not a pretty sight. Still, I marveled at what it revealed about what had taken place here. My 100-pound puma had stalked and killed this 800-pound elk, a mismatch extreme for nearly any carnivore. The evolutionary history of successful carnivores is written in their continual attempts to perfect methods to overcome diverse prey animals, just as that of prey is written in their attempts to escape predation. Yet despite all her skill and power, this cat probably failed to kill more often than she succeeded. But in this tight canyon, all the conditions had been right. The elk had come here because there were grasses and rushes to eat, and because the canyon's tight confines offered some shelter from winter winds that constantly rake the ridge tops. The puma was here simply because she knew that elk frequented the area, and because the canyon provided just the sort of terrain and cover she needed for a successful hunt. This was no random encounter of predator and prey.

No new snow had fallen, so I could follow the still-visible tracks etched in old snow. Hidden by brush, the cat had approached at an angle from higher up along the cliff base. Then she had bounded in, caught, and killed the elk with a bite to the neck. She later pulled his massive body, antlers and all, some 60 feet to this cover of fir tree and brush. She had fed, off and on, over the past two days, staying close to the kill to protect it from scavengers. She seemed to be alone; pumas willingly share their kills only with their young. I predicted she would be back shortly to eat again. An elk this size could feed her for two weeks or more.

Finding prey and foiling foes are the lot of any carnivore. Failure to do so means death for the hunter. Consider all the challenges such a cat must overcome in order to find, catch, kill, and eat an elk in the Salmon River Mountains of central Idaho, in winter. The tools for catching and killing are pointed canine teeth set in short, powerful jaws and toes tipped with retractable, jackknife-like claws. Pumas are powerfully built, with hind legs longer than front ones, giving them great prowess at jumping. Ample flexibility in the skeleton enables a wide range of movements to effectively employ the animal's deadly tools. As hunters of many different animals, pumas must have the capacity to learn and be versatile. The evolution of predators and prey can be traced in steadily increasing brain sizes for both types of animals. Indeed, elk and puma are closely matched, in other ways as well as relative brain-to-body sizes. Pumas eat flesh, which is much more easily digested than the twigs and grass that an elk must contend with. This is reflected in the puma's simple digestive system and the elk's much more complex, chambered

previous pages: **Alaskan brown bear**

stomach. Penetrating the tough hide of an elk and cutting off bite-size chunks of muscle take special tools, possessed only by carnivores: the so-called carnassial teeth, set back in the jaws, which function much like scissors.

Members of the order Carnivora—literally, "eaters of flesh"—are not the only mammals to eat meat. Virginia opossums, frog-catching bats, squid-eating sperm whales, and many other, very different mammalian species all share a taste for flesh. What makes some meat-eaters Carnivores—capital C—yet eliminates others? The flat, shear-like blades on the fourth premolars of the upper jaw and the first molars of the lower jaw define the group. It is this trait that unites the puma with animals as varied as tiny least weasels, bone-crushing hyenas, and clam-eating sea otters. Exceptions exist: In some Carnivora, such as modern bears and raccoons, these teeth have lost their carnassial shape during evolution.

All modern families of carnivores evolved from a primitive group called miacids, which resembled today's catlike civets and mongooses in that they had long bodies, long tails, short flexible limbs, and relatively small brains. Their wide paws had spreading digits tipped with sharp claws; they probably spent much of their time in trees. There have been many variations on this basic theme. Most modern carnivore families can be traced to a burst of evolutionary activity and diversification that marked the beginning of the Oligocene, 35 to 40 million years ago. Two major lineages of carnivores were present by then: The more doglike forms gave rise not only to today's dogs but also to bears, raccoons, weasels, seals, and sea lions; the catlike forms became modern cats, hyenas, mongooses, and civets. (Some taxonomists place seagoing carnivores such as seals and sea lions in a separate order, Pinnipedia, discussed in the following chapter.)

Eating meat is a luxury because it is so easily digested, yet it can be difficult to find and dangerous to catch. So it is that many carnivores seek other foods as well, their teeth and digestive systems adapting to the fare. In raccoons and bears, for example, the carnassial and other teeth have been modified so that they can grind fruits and vegetables as well as flesh. The puma, however, remains what we call a "hypercarnivore," because its diet consists almost entirely of the flesh of other vertebrates. It has a gut length about four times its body length. The red fox—labelled "omnivorous" because it eats fruits, nuts, and insects in addition to vertebrate prey—has a gut about five times its body length, as well as a caecum—an extension of the large intestine—to help process such foods. Cats, weasels, and mongooses generally lack a well-developed caecum. Bearlike carnivores include the bamboo-eating giant panda, the insectivorous sloth bear, the seal-eating polar bear, and the basically omnivorous black and brown bears—all with slightly modified dentition to suit their diets. The diversity of carnivores is tightly linked to their ability to occupy new food niches.

Worldwide, the order Carnivora embraces 246 living species, not so many when you consider that 977 species of bats and 2,052 rodent species share the same globe. Almost as many, in fact, as the world's 260 or so living primates. Only 40 carnivore species—about 16 percent of the worldwide figure—occur in North America north of Mexico. One, the puma, also occurs through Central and South America, giving it one of the largest geographical ranges of any mammal. Gray wolves, arctic foxes, brown and polar bears, wolverines, ermines, sea otters, and least weasels are all North American species that have ranged extensively into Eurasia, having what we call a Holarctic distribution. The river otters and lynxes of Eurasia and North America are two examples of closely related "sibling species" that share common origins and many common traits, but remain separate species. Likewise, the Siberian polecat has its parallel in the North American black-footed ferret. Such similarities help explain why North America's raccoon and mink were so successful at colonizing Eurasia, once humans put them there. The red fox of North America and Eurasia has been successfully introduced nearly worldwide.

All these species have adapted to life in widely fluctuating climates. They have physiological mechanisms that allow them to be either climate generalists or cold-adapted. Most of the 18 species in the Procyonidae—raccoons, coatis, olingos, ringtails, and kinkajous—live in climatically stable tropical and subtropical forests of the New World. They are warm-adapted, with low metabolic rates and relatively thin pelts. They lack a well-defined molt cycle and seasonal periods of fattening. Many are nocturnal, with only modestly diverse diets consisting mostly of fruit and insects. The familiar raccoon is, in fact, atypical of this family, which explains why it has moved beyond the tropics and subtropics. Raccoons have a relatively high metabolic rate, they undergo seasonal changes in body fat—a means to cool off in hot weather and retain heat in cold weather—and they enjoy an extraordinarily diverse diet. These traits enabled the raccoon to break out of the tropical mold that constrains the rest of the family and expand throughout the temperate regions of North America.

Diversity, of course, is largely a product of different climatic and regional regimes. Across the southern U.S. and northern Mexico, for example, great variation in topography and habitats have nurtured richly diverse carnivore species. Climate generalists such as the puma, gray fox, and long-tailed weasel overlap with cold-adapted species—lynx and least weasel—living at the southern extent of their ranges, as well as with warm-adapted species at the northern limits of their climatic tolerance, such as the ocelot, hog-nosed skunk, white-nosed coati, and ringtail.

We often think of carnivores as living in groups—a pride of lions, a pack of wolves. Yet beyond the bond of mother and young, group living actually is the exception rather than the rule. Most carnivores pursue a solitary existence, although they remain in contact with others through their calls and scent marks. Why? For one thing, some foods are more easily shared than others. Some are best obtained by cooperative hunting, as wolves do in winter, stalking large prey such as elk and moose. But for some species, group hunting would not be cost-effective. The general pattern among solitary carnivores such as cats and weasels is that adult females live alone (or with their young) on home ranges they know well. They focus on rearing their offspring without help from mates, and they may defend all or just part of their home range. Male home ranges usually overlap several female ranges. Males also defend their territories, depending on the season and the availablity of food. Their focus is to mate, not to rear the young. But among canids—wolves, foxes, and the coyote—group living is the norm, with males helping to rear litters. In fact, older offspring may remain with parents and assist with the raising of young. Such groups mutually defend their home ranges, at least part of the year. Size of the group depends on the availability of prey.

White-nosed coatis living in the Southwest illustrate yet another example of group living. Related and unrelated females of this species travel and forage in small groups. They do so not to make rearing or food gathering more efficient, but to protect their young from depredation by male coatis and other predators. Solitary males of the species, called coatimundi, live and travel alone; they will attack all other males, adult or adolescent, if given the chance.

Living at or near the top of the energy pyramid, large carnivores rarely occur in large numbers. So it is that they often are susceptible to human interference. Some, such as the grizzly bear, are highly endangered in the contiguous United States. Big, dangerous carnivores—the wolves, grizzlies, pumas—and fur-bearing carnivores became natural targets as humankind explored and exploited the continent. Yet in the historic period, only one North American carnivore has become extinct: the sea mink. This fish-loving species, once common among rocky coves and shores of Maine and eastern Canada, was extirpated by about 1880.

Some other carnivores have made more recent and remarkable turnarounds, including one I would not have thought possible when I was a graduate student studying pumas in Idaho's Salmon River Mountains. Thirty years ago, pumas lived only in our most remote western mountain ranges and in southern Florida (where they are known as Florida panthers). Government-sanctioned hunting and poisoning campaigns had kept puma numbers low, eliminating them even from Yellowstone National Park early in this century. Today, however, pumas are widespread in the West, even turning up in backyards and campgrounds. This is due largely to the creature's very flexible nature, and to the fact that it now has more deer to eat and more legal protection. What is even more remarkable to me today is that, both in Yellowstone and in those same Idaho canyons where I once tracked pumas, wolves now roam once again. Soon there may also be grizzlies along the Salmon River and in other parts of the West where they have been absent for a century or more. The re-establishment of complete predator-prey assemblages in some of our western mountains may be the beginning of a major conservation success story. Are we ready to accommodate the needs of these animals?

All of us who work, live, and play in such areas must learn tolerance and how to live with wildlife. Pumas, wolves, and grizzlies are magnificent hunters. Sighting one or even just knowing they are present somewhere nearby gives a thrill and a sense of completeness. Truly, our lives are richer now that they are coming back. ▪

BY JOHN SEIDENSTICKER AND SUSAN LUMPKIN

"**M**an's best friend," we call the domesticated member of this, the dog family. Whatever prompted humans to first take in carnivores as dangerous as wolves and set them on the path to domesticity remains shrouded in the ancient past, at least 12,000 years ago according to the fossil evidence. Less puzzling is why wolves acquiesced to such treatment. Highly social and intelligent, wolves live in groups with strict dominance relationships; the youngest animals occupy the bottom of the heap and quickly learn to obey superiors. Undoubtedly, people took in young wolves, assuming and never relinquishing their superior role.

Like wolves, most canids live in social groups, although group size can vary from a monogamous pair to packs of 20 or more members. They are generally slender and long-legged, with long muzzles and bushy tails, built for running down prey over long distances in open country. Canids possess keen senses of hearing and smell, and many follow scents like bloodhounds to find prey. Largely carnivorous, many will turn to fruit and insects if meatier prey is scarce. Group hunting enables some members of this family, such as wolves and African wild dogs, to kill prey much larger than they are. But other group-living canids, such as arctic foxes, hunt alone and stick to smaller prey. Canids shake a small mammal to break its back, while larger prey is bitten repeatedly on the legs and nose until its falters, whereupon the dog or dogs begin tearing it apart alive.

Of the world's 35 species of wild canids, 9 inhabit North America: gray and red wolves, coyotes, and red, gray, swift, kit, arctic, and island gray foxes. All but gray wolves and red and arctic foxes exist only in North America.

Gray Wolf (*Canis lupus*) No sound sends shivers up the human spine like the haunting howls of gray wolves. But these animals have little interest in how we respond; their howling is meant for the ears of other wolves, and it captures their attention far and wide. When all eight or more members of a wolf pack join in, they can be heard as far as six miles away. At close range, wolves communicate with whimpers, growls, barks, and squeals. Group howling tells other packs to stay clear of the home pack's territory. Howling also helps lost wolves find their packs following separation that might occur during prolonged chasing of prey; when all members are reunited, they often howl some more, in apparent celebration. Young wolves join the choir at two to three months of age.

A mated pair and their young of various ages make up a wolf pack; hunting and eating occupy most of the pack's waking hours. Group hunting enables wolves to take down large prey such as moose, deer, and caribou. But when it comes to eating, it's every wolf for itself. Mature adults weigh in at 75 to 110 pounds, and each can bolt almost 20 pounds of meat in a single meal. They hold back, however, when the pack's breeding female is caring for young, not-yet-mobile cubs. All pack members help feed both mother and young until they can travel with the group. Once ranging all over North America, gray wolves became restricted to Alaska, Canada, and Minnesota but have recently been reintroduced to several western states.

Howling gray wolves

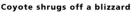
Coyote shrugs off a blizzard

Coyote *(Canis latrans)* Snow doesn't bother coyotes, which thrive in remote stretches of Alaskan wilderness or on the edges of Los Angeles—and almost any other habitat in between. Flexibility—in habitat, diet, and social groupings—characterizes these wiry canids, which vary from 20 to 50 pounds. Coyotes eat fruit and rodents as well as pronghorn fawns, deer, and carrion. When existing on small food items, coyotes live alone; when large prey abounds, they form packs.

The one thing these determined survivors seem unable to cope with is wolves; wolves kill them. The decline of wolves that began with Europe's colonization of North America created a vacuum that coyotes readily filled, though they also suffered persecution from humans. Today, this species occupies the entire continental U.S. and Canada up to tree line, as well as Mexico and Central America. However, recent reintroductions and natural expansion of wolf populations may change the coyote's luck. In the few years since wolves were returned to Yellowstone National Park, coyote numbers there have declined dramatically.

Mexican wolves stand watch

Mexican Wolf *(Canis lupus baileyi)* At 50 to 90 pounds, this slightly smaller subspecies of the gray wolf—locally known as *el lobo*—once lorded over Mexico and the southwestern United States. From the late 1970s to 1998, however, it roamed no farther than the confines of zoo exhibits. Like the gray wolf and the red wolf *(Canis rufus)* of the Southwest, it was trapped, shot, and poisoned into near-extinction. Both red and Mexican varieties survived only in zoos, while small populations of grays held on in wilder areas of northern Montana, Minnesota, and Michigan. But in 1998, to the cheers of conservationists and jeers of ranchers, three Mexican wolves were released in Arizona's Apache National Forest. Similar reintroductions—a 1987 effort with red wolves in North Carolina and the 1995 release of gray wolves in Montana, Wyoming, and Idaho—have proved successful. About 100 red wolves now roam North Carolina and Tennessee, while 160 gray wolves inhabit their reintroduction areas; these populations should continue to grow. Perhaps the Southwest will one day boast similar numbers of its native wolf.

Arctic Fox (*Alopex lagopus*) Battling the brutal cold of the far north requires strong defenses, and arctic foxes have largely won that fight. Smaller than the closely related red fox, they also possess shorter ears, legs, and tails—features that help reduce heat loss. Their winter coats of fur are so warm that these animals don't begin to shiver until the thermometer hits 94° F below zero.

But keeping warm is only half of the challenge of living at these latitudes; finding enough to eat during the long Arctic winter is the other. Arctic foxes rely largely on lemmings, which are both summer staple and winter fare, even though these small rodents disappear in burrows beneath winter snows. Equipped with discerning senses of sound and smell, the fox detects lemmings under the snow cover and rapidly digs them out with its front feet. Arctic foxes also scavenge the carcasses of caribou and marine mammals, and sometimes hunt seal pups, which, like lemmings, can be sniffed out and dug up from the snow surface.

Red fox sniffs a tree and watches

Red Fox *(Vulpes vulpes)* Unable to climb as gray foxes do, the red fox can only look longingly at bird nests and arboreal prey. But it can leap astonishing distances, jumping as far as 16 feet to nab a mouse. It weighs 8 to 15 pounds and grows up to 43 inches long, one-third of which consists of its bushy, trademark tail. Sometimes referred to as the catlike canid, red foxes possess long legs, large ears, and remarkable abilities to detect hidden rodents by listening for their movements, then springing and pouncing, pinning them with their paws.

Like most canids, red foxes enjoy a rich social life. They live in mated pairs, sometimes accompanied by older offspring that help care for the five or so cubs born each spring. Unlike wolves, however, they have been highly successful in the modern world. Omnivorous and opportunistic, they currently inhabit most of the Northern Hemisphere and have been naturalized in Australia. They coexist among people in cities and suburbs, and have expanded both in population and range.

Family Ursidae **Bears**

Largest of all land-living carnivores, bears possess heavy bodies, powerful limbs that end in long, sharp claws, and huge heads with piercing canine teeth. They run fast over short distances on the soles of their broad feet and boast an acute sense of smell. They eat meat and fish, and do not disdain carrion. Some take young hoofed mammals and other easily caught prey. Yet all but the polar bear have become omnivorous, subsisting as well on diverse foods that range from lichens to roots, nuts, berries, seaweed, honey, grubs, and ants. This dietary regime is reflected in their broad, flat molars, designed for chewing vegetable matter. The key to a bear's survival is finding enough food to satisfy the energy demands of its large size.

To do this, bears have become problem solvers with excellent memories for finding aggregations of food that are widely separated in space and that appear at different times of the year. They range over huge expanses of territory and remember details in their quest. Still, bears exist on the edge of their energy needs. They must fatten up enormously in the fall, when food is abundant, then virtually shut down activity for the food-scarce winter months. During this time, female bears give birth to tiny babies; at only 1 to 3 percent of the mother's weight, they rank as the smallest, relatively, of all carnivore newborns.

People have long been fascinated by bears' annual cycles of disappearance and re-emergence, as well as by the awesome power, acute intelligence, and perseverance of these creatures. Indeed, bears have become totemic symbols in all human cultures that have lived in their midst. Three of the world's eight species of ursids inhabit various parts of North America: polar, grizzly, and black bears. Only the black bear lives exclusively in North America.

Black Bear (*Ursus americanus*) Venturing out of its usual winter seclusion, this black bear won't stay exposed for long. An elusive creature of the forest, it has managed to remain a virtual ghost in human-dominated landscapes—an ability that has served it well. After centuries of persecution and decline over much of its range, it now is holding its own or increasing in areas where it managed to survive, principally the boreal forest and major mountain ranges of the continent; some people have stumbled across them in suburban backyards.

Generally preferring to avoid humans, a mother black bear with cubs can be dangerous—and scrambling up a tree won't protect you, as black bears are agile climbers. Indeed, females often park their cubs in trees while they go off to forage.

This smallest of North American bears ranges from 200 to 600 pounds for adults, with females at the small end. Its diet changes with the seasons. Spring menus often consist of grasses and green plants. Cherries, elderberries, blueberries, and other fruit provide summer fare. In the fall, black bears seek high-energy nuts such as acorns and walnuts, as well as grapes and other fall fruits. Abundant food in the fall is critical to preparing these animals for the lean winter months. During two autumn months, black bears normally gain 1 to 3 pounds every day. This 60 to 180 pounds of fat will carry them through four or five months of winter fasting.

Black bear in Montana

Brown or Grizzly Bear

(*Ursus arctos*) Learning to be a grizzly takes time. Cubs—usually twins but sometimes singles or triplets—spend at least two years following their mother far and wide as they learn what to eat, where to find it at various times of the year, where to find safe denning sites, and how to protect themselves in the open habitats in which they live. Following the principle that the best defense is a good offense, cubs practice charging and rear up on their hind legs to appear larger than they are. By the time they are ready to leave their mother, the family makes a formidable team, unlikely to be challenged by any other bear, even at feeding grounds.

The term "grizzly" stems from the gray-tipped fur found in some members of this species, which tend to live inland and average 600 to 800 pounds. Brown bears inhabit coastal areas and can grow twice as massive. Once found across Eurasia and western North America, the species survives mostly in northern Asia, Alaska, and western Canada, with isolated populations in Europe and the Rocky Mountains of the western U.S. This bear's color can be confusing; both grizzlies and brown bears vary from almost white through shades of brown to black, and can change through the year. American black bears vary from blonde to cinnamon, chocolate, or black. To tell the species apart, look at the features, not the color. A grizzly's face is dish-shaped, while a black bear's muzzle is aquiline. Grizzlies also display a prominent hump above the shoulders, while black bears do not.

Mother and cubs browse for berries

Alaskan brown bear, alert to its surroundings

Pair of polar bears play-fight

Polar Bear *(Ursus maritimus)* Armed with slashing, saberlike claws, with prominent canine teeth, and with a body weight of up to 1,700 or so pounds, male polar bears pack considerable punch. Real combat between mature males—which does occur in spring when they compete for breeding rights—can leave the snow blood red. But in the autumn these powerful animals engage in more gentlemanly jousts that usually leave neither combatant the worse for wear. The reasons for this remain mysterious, although scientists speculate that such play-fighting may provide practice for the more serious contests of the spring breeding season.

First appearing only 70,000 years ago, polar bears evolved from brown bears. Their white fur, with none of the color variation seen in brown bears, blends perfectly with their inhospitable Arctic habitat. Abandoning the omnivorousness of their cousins, polar bears became mostly carnivorous, learning to exploit the riches of the sea and of seals in particular. Hunting seals requires reading the ice for places where they come up for air. Once a promising crack is spotted, a polar bear patiently waits beside it, sometimes not moving a muscle for hours. Finally an unsuspecting seal pops up for a breath—and the bear springs, grabbing it with massive paws before it can dive back to safety.

Monarch of ice and water

Family Procyonidae Procyonids

From a two-pound ringtail with its big, sad-looking eyes to a 20-pound raccoon, the 18 species of the family Procyonidae share one prominent defining feature. All possess highly mobile, five-fingered hands, with which they rummage, dig, and manipulate items from insects and fruit to garbage can lids and corn stalks. They also have pointed muzzles, broad crushing molars, forward-set eyes, and plantigrade feet. Agile and acrobatic, procyonids easily climb to escape their predators, and they bear and rear their young in trees. The tropical members of this exclusively New World group depend on fruit, but those living in temperate zones eat whatever they can find in abundance. Although scientists once considered Asia's giant panda to be a procyonid, most now classify it as a bear.

Raccoon harvests rose hips

Raccoon *(Procyon lotor)* Fingers reaching out to pluck the fruit of wild rose bushes, this "masked bandit" reveals one reason for its success: It makes the most of whatever foods are abundant, from mulberries and crayfish to ripe corn in a field and scraps from your garbage sack. Their diverse diet, coupled with an ability to fatten up in the fall in preparation for winter, helped raccoons to expand northward from the confines of Central America's tropical forests. They live throughout the United States, reaching around 20 pounds and 32 inches in total length. Greatly reduced numbers of natural predators such as wolves and pumas have enabled raccoons to become superabundant in some places, which has meant trouble for species such as sea turtles and songbirds, whose eggs make tasty, nutritious meals. Generally harmless to humans, raccoons may spread rabies in some areas.

Wide-eyed ringtail

Ringtail *(Bassariscus astutus)* For this species, hanging out on a limb is as natural as leaping along rocky outcrops or dashing headlong down a tree trunk. Sleeping by day and foraging by night, ringtails are the most carnivorous members of the procyonid family, though they also enjoy persimmons and other sweet treats. The two-pound creatures rely on both speed and stealth; they pounce on big prey such as a pocket gopher, killing it with a sharp bite to the neck, and can expertly grab a grasshopper in their dexterous fingers. Ringtails grow nearly three feet long—half of which is tail—and range from Oregon to southern Mexico and as far east as Arkansas, occupying diverse habitats with long dry summers and moderate winters. When threatened, they fluff out their conspicuous black-and-white namesake tails and emit a pungent musk—which has led to the nickname, "civet cat." Their main predators are night-hunting great horned owls and snakes.

C o a t i *(Nasua narica)* Banded tail held high as it scurries along the forest floor, the coati—twice the size of a ringtail—seems a bundle of restless energy. It pokes and probes with its long mobile snout and sharp claws, seeking grubs, lizards, fallen fruit, and other tasty morsels among the leaf litter, under fallen logs, and in any hole or crevice it comes across. Group-living females, ever whirring, clicking, and woofing to keep in contact with each other, appear especially high-strung. Lone males, once believed to form a totally different species called coatimundis, make far less noise. Because males sometimes kill the young, female coatis actively drive them away from their groups, tolerating their presence only during the breeding season. Living from New Mexico and Arizona well into South America, coatis are active in daytime, perhaps to avoid the multitude of nocturnal predators such as ocelots and jaguars, with which they share their habitat.

Family Mustelidae Mustelids

This family has inspired some of the nastiest animal metaphors in the English language. No one wants to be called a skunk, or to be accused of weaseling out. The words badger, ferret, and polecat all evoke negative images. Why such negativity, from which only otters are spared? A diverse group of about 65 species worldwide, 15 of them North American, mustelids range in size from 1-ounce least weasels to 60- to 100-pound sea otters.

All possess great strength, especially bite strength, for their size. Most kill prey far larger than themselves; many, like the wolverine, appear fearless and aggressive. Yet their torpedo-shaped bodies and high-speed locomotion make them escape artists as well, able to duck into the smallest holes. That same ability helps them hunt burrowing prey. Small size and high metabolic rates demand that mustelids hunt and eat almost constantly. When they locate a food bonanza, say a hen house, some species kill every last hen—not from bloodthirsty rage, but to set aside future meals. Their senses of smell and hearing are acute, but their eyes are small. Most

previous pages: **Short-tailed weasel takes a ruffed grouse**

mustelids produce musk from anal glands, used primarily for scent communication and only secondarily for defense, as skunks do. All in all, they compose a most fascinating group in need of improved public relations.

Long-tailed Weasel *(Mustela frenata)* Ranging from southern Canada to northern South America, this species varies in color with latitude. More northerly individuals turn snowy white in winter, like the ermine. Those of the mid-latitudes—where snow is not everlasting—are chestnut brown mottled with white. In the south, where snow rarely falls, fur color remains brown year-round.

This largest of North American weasels grows up to 9 ounces and up to 22 inches long. Least weasels—sometimes called snow weasels—exhibit similar coloration patterns but weigh only 1 or 2 ounces and rank as the world's smallest carnivore. Despite their tiny size, they can be ferocious predators. Dashing at prey as large as rabbits, they often seize their victim by the nape and bite through its skull in a swift, efficient motion. A kill may take as little as ten seconds.

Fisher prowls an aspen

Fisher *(Martes pennanti)* Despite its name, the fisher does not fish. This arboreal speedster has learned to crack one of nature's best defenses: a porcupine's barbed quills. Porcupines can make up as much as a third of a fisher's diet, though they usually dwarf their 3- to 12-pound antagonist. Slim and low to the ground, the fisher aims for the porcupine's face—the one area of its bristling body that is both vulnerable and accessible. It is fast and agile, the porcupine slow and clumsy. Often the two circle one another, battling until the fisher's darting feints and bites take their toll. Combat may last half an hour, but finally the porcupine collapses, and the fisher flips it over and rips into its soft belly. For the fisher the spoils are well worth the prolonged effort; a big porcupine can feed it for two weeks. Surprisingly, while most fishers bear quill scars, they seem unaffected by them—unlike most other mammals. Athough viewed as porcupine specialists, fishers rely mainly on snowshoe hares, which are staples when porcupines are in short supply.

Striped Skunk *(Mephitis mephitis)* If its conspicuous black-and-white coloration isn't enough of an early warning to an intruder, a striped skunk's fluffed-out tail signals its odiferous intentions. Back off, it says, or smell the consequences. If an intruder ignores this message, the skunk issues another. It whirls around and performs a handstand to offer a whiff of the pungent musk from its anal glands. If that's still not enough, the skunk drops to all fours and blasts the foolish creature's face with a spray of stinging, stinking musk. Eyes burning, nearly blind, the intruder gets the word: skunks are not to be trifled with. In fact, very few predators take them on more than once. Night-hunting great horned owls are a significant exception, but then birds possess a notoriously poor sense of smell.

Striped skunk sends a message

River otter skips through the shallows

River Otter *(Lontra canadensis)* No animal seems to have as much fun as a river otter. These free spirits of the weasel family often swim, dive, and play with "toy" stones and pebbles. Weighing up to 30 pounds, otters are agile and quick, with a reputation for high intelligence. Truly amphibious, they inhabit freshwater estuaries, streams, rivers, and lakes, as well as the banks of these waters.

Growing perhaps four feet long, they usually stalk slow-moving fish and crustaceans, also birds (and their eggs), amphibians, insects, and mammals. Their highly sensitive facial whiskers help them detect and catch prey even in murky waters. They swim underneath waterbirds and grab them from below; they scrabble around in mud- or rock-bottomed ponds, using their dexterous front paws to capture crayfish and other tidbits; occasionally they also will hunt birds and mammals on land. Male river otters live alone, joining females only during the

Sea otter feasts on crab

annual breeding season. Females enjoy the continuous company of their cubs, which they teach important skills, such as how to locate and catch food in water. River otters of both sexes and all ages, however, may meet without strife.

Sea Otter *(Enhydra lutris)* Floating languidly in the sea, wearing fabulous fur coats, grazing on succulent crabmeat or abalone, often ogled by cooing human fans, sea otters seem to enjoy California's good life. It wasn't always so, of course. During the 18th and 19th centuries they were hunted so ruthlessly for their fur that their numbers fell from about 200,000 in the mid-1700s to fewer than 2,000 in 1911, when an international treaty finally ended the commercial harvest. Slowly, they have recovered. Today about 100,000 exist, living in all but a quarter of their original range, which extended from northern Japan through the northern Pacific rim to Mexico. They grow to more than 6 feet in length and can weigh nearly 100 pounds.

Some 2,000 individuals, considered genetically distinct from populations in Alaska and Asia, range along coastal California. And here, curiously, their recovery has largely stalled. For reasons that are poorly known, 40 to 50 percent of sea otter newborns die before they are weaned, and many adults die before they get old. Moreover, many commercial fishermen begrudge the sea otters taking food that they can so profitably sell to human seafood lovers. For these otters, at least, the California good life seems as illusory as it is for so many humans.

Family Felidae **Cats**

Powerful predators designed to hunt and kill with uncommon stealth, cats capture the human imagination like no other carnivore. Our ancient ancestors must have envied the predatory prowess of felines, their admiration mixed with fear. From the biggest to the smallest, all 36 species of cats—7 of them North American—possess similar features. Their teeth, set in strong jaws, are highly specialized: Large canines grab and kill prey; scissor-like molars, called carnassials, tear chunks of meat from a carcass; small incisors scrape the last scraps from bones.

Well-muscled forelimbs connect strong shoulders to sharp claws. Cats see and hear well, possessing both binocular and color vision. They move silently on well-padded paws, and run with speed and agility. All but lions and cheetahs are solitary hunters, stalking or waiting in ambush, then charging and dragging prey down unassisted, with their forelimbs. A crushing bite to the nape or throat kills the victim, often far larger than the hunter. Cats rely almost entirely on vertebrate prey, and are purely carnivorous, eating no vegetable matter at all. Young felines usually remain with their mothers only one to two years. Living alone does not make cats asocial, however, for they remain in relatively constant contact with others of their species, through scent marking and—sometimes—vocalizations.

Until about 12,000 years ago, our continent boasted an assemblage that included lions, cheetahs, pumas, and huge jaguars, not to mention saber-toothed cats. Today, jaguars hang on tenuously in North America, while the continent-wide range of pumas has been cut in half. In addition to these two large cats, the bobcat, lynx, margay, ocelot, and jaguarundi also live in North America.

Jaguar *(Panthera onca)* The sleepy look of this preening jaguar masks its extraordinary strength. Built like a heavyweight wrestler, this largest of American cats weighs in at up to 250 pounds, with short, thick legs, large feet, and a large head. Its powerful bite can pierce even the armor of turtles and tortoises. Mostly creatures of the forest, where big prey are thinly distributed, jaguars are not picky eaters. They prefer tapirs (large mammals related to horses and rhinos), but make do with peccaries, capybara, deer, monkeys, squirrels, armadillos, small birds, fish, and reptiles. Their penchant for horses and cows has resulted in conflicts with ranchers, who have largely exterminated jaguars from their vicinity. Hunting for fur has further reduced their numbers.

Primarily a South and Central American species, the jaguar is increasingly rare in Mexico, where perhaps only 500 remain; it became extinct as a breeding population in the southwestern U.S. early in this century. About 60 jaguars have been reported in Arizona and New Mexico since the 1890s, and almost all were killed. Most recently, in 1996, a hunter tracked—and didn't kill—one jaguar in Arizona, while at least one other was seen there that same year. Perhaps young males in search of new territory, these appearances raise the possibility that jaguars might return to the U.S. Successful recolonization will depend, however, on the good will of hunters and ranchers.

Jaguar cleans its fur

Closing in on a mule deer

Mountain Lion (*Puma concolor*) Puma, mountain lion, cougar, panther, painter, and catamount—these and many other names refer to this cat, which lives from the Canadian north to the southern tip of South America. Biologists have long puzzled over the puma's taxonomic place, and have given it almost as many scientific names as it has common ones. Conventional wisdom once divided cats into two main groups, big and little. Lions, tigers, jaguars, and all types of leopards were big cats; all the rest, except for the greyhound-like cheetah, were small cats. Although male pumas can weigh more than 225 pounds and measure 9 feet long— thus they are larger than leopards—they were put into the small-cat group, largely due to their skulls. Experts referred to them as having small-cat heads on big-cat bodies. But more recently, scientists have gone beyond overt physical features such

Puma on the alert

as bones and teeth and looked at DNA to determine interspecies relationships. It turns out that pumas' nearest living relations are none other than those "aberrant" doglike cheetahs and another unusual feline, the weasel-like jaguarundi. So we now have a "puma lineage," placed within an expanded big-cat grouping.

Despite the apparent superficiality of classifying cats by size, this practice does help us understand and explain their behavior. Large lions and tigers, virtual twins under the skin, depend entirely on big, substantial prey and may go days between big, gorging meals. Small cats, such as margays, jaguarundis, and ocelots, hunt smaller prey such as birds and mice, and so must hunt almost constantly. Cats of intermediate size, including pumas, jaguars, and lynx, can kill large deer but also can survive on smaller prey. Their hunting frequency varies accordingly.

Bobcat up a tree

Bobcat (*Lynx rufus*) Just as basketball legend Michael Jordan doesn't make every shot, feline hunters don't make every kill. Bobcats specialize in hunting rabbits and hares, but perhaps five out of six of their attempts end in failure. Upon sighting a snowshoe hare foraging in the snow, a bobcat will use every stone and shadow to conceal its approach. When it closes the gap to about 30 feet, it rushes to pounce. But the alarmed hare may jump first, eluding the cat's grasp, and the hunter must try again, and soon. On average, a 20- or 30-pound bobcat needs one bunny a day to survive. Shortfalls are made up by capturing large numbers of mice, rats, and squirrels, and occasionally taking birds, fish, and large insects. Bobcats also will go after meat bonanzas, such as adult deer, when opportunity knocks. With deer outweighing bobcats by a factor of seven or eight, the cats fare best when the deer have been stressed by winter food shortages and hampered by deep snow. Bedded-down adult deer and resting newborns also make fairly good targets.

The bobcat's dietary flexibility is mirrored in its habitat flexibility. Bobcats live from coast to coast and from northern Mexico to the U.S.-Canada border, equally at home in marshes or mountains, hot deserts or cold woods. Only intensive agriculture and clear-cut suburbs exclude them. This, coupled with the bobcat's ghostlike ability to hide itself from humans, contributes to its survival in the modern world, despite continued and extensive trapping for its luxuriant fur.

previous pages: **Bobcat pursues its favorite food, a snowshoe hare**

Lynx on the prowl

Lynx *(Lynx canadensis)* Making its home in northern coniferous forests that are covered much of the year in ice and snow, lynx are well known for large paws thickly padded with fur; they work just like snowshoes. The ability to travel long distances is essential to this species, which weighs up to 40 pounds and whose home range varies with the availability of food. When prey—that is, snowshoe hares—are abundant, a lynx's home range may be as small as two or three square miles. When prey are scarce, that range may expand to more than 90 square miles.

Apart from food considerations, successful male lynx include in their range the home ranges of two or more females with which to mate. Solitary males and females tolerate each other—albeit at safe distances—except during mating; both defend their territory from others of the same sex. Because suitable territories exist only in limited numbers, young males may have to travel long distances in hopes of finding a vacant one. Females often fare better: A mother lynx may let a daughter carve off a corner of hers. Whenever a territory holder dies, another lynx claims all or part of it within days, for a lynx without a place to call its own does not breed. It hunts in marginal areas and is subject to attacks from territory holders. Given this perspective, it is not surprising that males at times fight to the death to keep or gain a territory. Such a social system, with little variation, holds true for many cats around the world, from tigers to pumas to margays.

Pinnipeds

Order Pinnipedia

BY BURNEY LeBOEUF

Linnaeus, the 18th-century Swedish botanist and master classifier, considered seals a "dirty, curious, quarrelsome tribe, easily tamed and polygamous." Quarrelsome and polygamous are safe predictions for most mammals. Female seals quarrel to protect their pups; males fight for access to females in heat, attempting to mate with as many as possible and doing nothing to care for the young. As for being easily tamed, few seal species have been tried. Best known is the "circus seal," actually the California sea lion. Invariably females are used, because the far larger males are too expensive to maintain.

Seals, sea lions, and the walrus all belong to the order Pinnipedia—which means having feet that are either feather-, fin-, or web-like. They are also called "aquatic carnivores" because they share common roots with terrestrial carnivores. Their ancestors, primitive bearlike creatures, entered the sea 25 to 40 million years ago. Pinnipeds differ from terrestrial carnivores in that they are exclusively carnivorous, eating a variety of fish, squid, krill, and other animals, but never carrion. They also have lower reproductive potential, giving birth to single pups (rarely twins) once a year or every two years.

Pinnipeds differ from cetaceans—whales and dolphins—in their ancestry and their adaptations to marine life. Cetaceans, whose terrestrial ancestors were even-toed ungulates, took to the sea about 60 million years ago. While cetaceans and pinnipeds both forage at sea, only pinnipeds rely on land, ice, or sandbars to give birth to and nurse their pups. In adapting to the ocean, pinnipeds lost speed and agility on land, making them easy prey for land predators such as grizzly bears, wolves, and mountain lions. This is why they often prefer islands or ice, areas free from such predators; when on the mainland, they remain close to water's edge.

Pinnipeds comprise three families: true seals or phocids (19 species), eared seals or otariids—which include sea lions (5 species) and fur seals (9 species)—and, all by itself, the walrus, an odobenid. The walrus's size, prominent tusks, and thick skin make it easy to identify. Phocids have no external ears, pull themselves over the ground with a humping, caterpillar-like motion, and swim sinuously, much as fish do. In contrast, otariids have external ears, walk on all four limbs but as if the hind ones, brought under the body, are tied together. They swim using power strokes of their long, paddle-like forelimbs. In 1880, naturalist Joel Allen defined the two families as "wrigglers" and "walkers."

The current world population of pinnipeds has been estimated at about 50 million, 90 percent of which are phocids. Most are crabeater seals of the Antarctic (about 30 million), followed by arctic ringed seals (6 to 7 million) and harp seals (3 million). Pinnipeds generally inhabit polar to temperate waters, preferring food-rich areas such as coastal upwellings and along the edges of pack or fast ice. Only monk seals live in less productive tropical waters, and they are rare or endangered. The Caribbean monk seal became extinct in this century.

All but four of the Northern Hemisphere's pinnipeds occur in North American waters, the exceptions being Russia's landlocked Caspian and Baikal seals, and the

previous pages: **Walrus bask en masse on Round Island, southeastern Alaska**

Mediterranean and Hawaiian monk seals. North American species include California and Steller's sea lions, Alaska and Guadalupe fur seals, northern elephant seals, gray seals, common or harbor seals, walrus, and ice-loving species such as harp, ringed, hooded, bearded, and ribbon seals. They occur in three general areas: the West Coast, from Baja California to the Alaskan Panhandle; Alaska proper, including the Aleutian Islands and the Bering, Chukchi, and Beaufort Seas; and along the North Atlantic, from Baffin Island and Hudson Bay to the Gulf of St. Lawrence and—for a few species—farther south. The West Coast is inhabited by sea lions, fur seals, elephant seals, and harbor seals; Alaska by the northern fur seal, northern sea lion, walrus, bearded, harbor, ribbon, and ringed seals; northern Canada by walrus, ringed, harp, hooded, harbor, and gray seals. We know much about the animals living in temperate waters, but little about most of those that live near the ice.

Socially, pinnipeds fall into two camps. Some are highly gregarious, coming together in large aggregations at certain times of the year. Others are solitary. Sea lions, fur seals, elephant seals, and gray seals fit into the first category. Besides being highly social, they exhibit marked sexual dimorphism, with males up to five times larger than females, as well as extreme polygyny, a few males getting many mates and most males getting none. A northern fur seal bull defends a territory that may include as many as a hundred females with whom only he mates; male elephant seals form dominance hierarchies, power structures based on vocal threats and fighting. Such highly polygynous systems are characteristic of pinnipeds that breed on islands, natural sanctuaries from predation. Since islands are in limited supply, females generally give birth and rear their pups together, in congregations. This situation gives top-ranking males the opportunity to have multiple mates. One result is selection among males for traits such as large size, an obvious advantage in fighting. Of these gregarious pinniped species, most mate and molt on land.

In contrast, most seals that feed along the ice edge—ringed, spotted, ribbon, bearded, and hooded seals—are generally solitary. During breeding season, their groups consist of at most a female, her pup, and an attending male. Females have no need to crowd together, since the ice habitat is uniform and extensive. Males search for a lone female that is about to give birth or has just done so on the ice, wait for her to come into postpartum estrus, mate, and then seek other females. Polygyny is necessarily modest, because females are widely dispersed and the mating season is short. The lack of female clumping precludes the evolution of extreme polygyny and associated traits; males and females are similarly sized in all but the hooded seal. All of these ice-oriented seals mate in the water.

The walrus, as usual, is different. Though it mates in icy waters, it is moderately polygynous, the males being larger than females. During breeding season, females congregate on floating ice; swimming males compete around them, using visual and vocal displays to lure females into the water to mate.

For most pinnipeds, the period between mating and birthing lasts nearly a year. Fetal development takes only about nine months, but is preceded by an initial period of delayed implantation, which lasts two to three months. This delay enables females of breeding age to give birth at the same time of year, every year.

Pup-rearing strategies vary among different species. Walrus may be the only pinniped that uses the "aquatic nursing" strategy common to whales, dolphins, dugongs, manatees, and sea otters. A few days after birth, the mother walrus goes to sea to forage, accompanied by her pup. She nurses on the run, either in the water, on ice, or on land. In about five months the youngster begins to feed on its own, but lactation continues for about two years. Walrus milk contains some 26 percent fat, less than half that found in other pinniped milk, but far richer than cow's milk, which has around 4 percent fat.

Most mammals, including humans, eat more during lactation. Most phocid mothers, however, fast during a short, intense lactation before weaning their pups abruptly. They stay ashore throughout this period, nursing their pups several times a day while totally abstaining from food or water. Their milk, high in fat and energy, is produced entirely from body reserves stored prior to giving birth. It is so rich that the pup quickly doubles or even triples its birth weight, while the mother may lose 40 percent of her mass. Hooded seals illustrate this "fasting strategy" in the extreme; their four-day lactation period is the shortest of any mammal, yet a 50-pound newborn will gain 15 pounds per day on milk that is 61 percent fat. By far, most of the weight gained is deposited as fat. The daily energy intake of the pup relative to its body weight will be about 15 times greater than it is for a terrestrial mammal. A similar fasting strategy also occurs with gray and elephant seal mothers, whose lactation periods last 16 and 28 days, respectively.

Otariids, however, use a "foraging-cycle" strategy; females give birth on traditional rookeries, nurse pups from blubber stores for about a week, then go off to sea to eat for a couple of days to a couple of weeks. They alternate such foraging trips with onshore nursing visits of 1 to 3 days. While the fat content of otariid milk is somewhat lower than that of phocids, the lactation period is longer—four months to three years, depending on the species. Weaning is gradual; the pup begins to feed on its own while still suckling. In fact, sometimes a mother northern sea lion will nurse a daughter that is herself nursing her neonate. Small phocids, which cannot store much blubber, also use a foraging-cycle strategy.

Maternal strategies affect other aspects of life as well. Aquatic nursing works well for walrus, because the bivalves and benthic organisms on which they feed often occur in predictable locations. But species that use a foraging-cycle strategy must breed and pup relatively near a high concentration of prey. Also, they must expend great amounts of energy commuting to and from foraging areas, must limit their foraging time, and must recognize their pups when they return to nurse them. Fluctuations in the distribution and availability of prey, such as those associated with El Niño events, may require foraging mothers to range farther and farther from the rookery—while the hungry pup waits at home. During the severe 1983 El Niño, for example, all fur seal pups in the Galápagos Islands died of starvation, due to long foraging absences of their mothers. The 1998 El Niño was similarly devastating to California's pinniped rookeries on San Miguel Island.

Seals that fast while nursing face different challenges. At weaning, an elephant seal pup is nearly half fat, and can be so bloated that it will roll downhill. It has sufficient energy reserves to develop for one to three months, after which it will begin to feed on its own. This leaves the mother free to undertake two extensive foraging trips a year, for herself and her future offspring. She is not limited by how long it may take to find and catch a suitable amount of prey; she can exploit more dispersed or patchy food resources over months at sea—as long as she can produce large amounts of high-fat milk that next year's pup will need to survive.

Unlike terrestrial predators, marine mammals find, pursue, and capture their prey while holding their breath. This limits feeding depth and time, of course. The larger phocids dive deeper and longer than otariids or the walrus. Elephant seals are the ultimate diving machines. Males migrate north twice a year from rookeries in California and Mexico to foraging areas along the continental margin from Oregon across British Columbia and southern Alaska to the Aleutian Islands, a round-trip voyage of some 5,500 miles. Females meander widely in the northeastern Pacific, up to 3,800 miles from the California rookery, feeding along the way. They dive continually, day and night, during these two- to eight-month-long excursions. Dives last an average of 20 minutes, reach depths of 1,200 to 1,900 feet, and are followed by 2.5 minutes at the surface. But some individuals go as deep as a mile down, staying submerged for as long as 90 minutes. Such performances exceed those of most whales.

The diving pattern of eared seals is known almost exclusively from studies of lactating females, because the diving instruments used are most easily recovered when the female returns to nurse her pup. Foraging trips of lactating California sea lions last two or three days and may average 130 miles. According to one study, 41 percent of that time was spent commuting to and from feeding areas, 33 percent diving for food, and only 3 percent resting. Otariids usually dive 100 to 200 feet deep and average less than 2.5 minutes under water. The largest otariids, however, may dive more than twice as deep and stay down six times as long.

I'm still not sure why Linnaeus considered seals curious and dirty. If he saw seals in his native Sweden, they were most likely harbor seals, and such adjectives just don't fit. Although harbor seals live fairly near humans in bays and estuaries, they remain shy, ever alert, and difficult to approach. The most that a fisherman or beachcomber sees of them is a brief head-poke above the water as they quickly survey the intruder. Today's biologists are most impressed by other characteristics of pinnipeds: their worldwide distribution across diverse climates, their ability to transfer enormous amounts of milk energy to their pups, their capacity for fasting while breeding, their long foraging migrations, and their proficiency at finding food in the dark and cold ocean, while holding their breath and experiencing pressure changes that would kill any terrestrial mammal—including us. ■

Family Otariidae Eared Seals

External ears, sharp and conical teeth, and hairless hind flippers that can be brought forward under the body to aid movements on land mark all members of this family. They are a uniform group in several respects. Males of all species are polygynous, having multiple mates. They also are larger than females—three or four times as big, in some species—and are territorial to varying degrees. Females bond closely with their pups for several months, until the young are weaned.

The family consists of 14 species, 4 of which occur in the coastal waters of North America; they are grouped into two subfamilies: fur seals and sea lions. Fur seals are distinguished by their smaller size, more pointed snout, and luxurious coat of densely packed underfur, which keeps them warm and waterproof. Millions of fur seals have been harvested for their valuable pelts during the last 200 years. One source estimates that three million northern fur seals alone were slaughtered between 1786 and 1828.

The Guadalupe fur seal, which once bred along the coasts of Mexico and southern California, was nearly exterminated in the last century. Today, it remains rare and endangered, breeding only on a single oceanic island off Mexico and numbering fewer than 4,000 individuals. In general, sea lions were not hunted as intensively because their pelts were far less valuable, being used mainly for the making of glue. From about 1865 to 1930, however, sea lion populations were decimated as a result of the wasteful, destructive "trimmings trade." Breeding-age males were killed for their genitalia, gall bladders, and whiskers; the organs were dried and sold in Asian markets for their presumed aphrodisiacal and medicinal powers, while the whiskers were used to clean opium pipes and to pick teeth.

Many members of this family exhibit delayed implantation: After undergoing early stages of development, the fertilized egg then ceases growth for weeks or even months before attaching to the uterine wall and resuming development.

California Sea Lion (*Zalophus californianus*) Of all pinnipeds, the California sea lion may best exemplify the conflicts that can develop between people and wild animals. This species, the most common pinniped in aquaria and oceanaria, is known and loved around the world as the "circus seal" that readily balances a ball on its nose. It is playful and gregarious, also adaptable and highly intelligent. The latter traits have helped it become a nuisance and even a scourge to sportsfishermen in search of salmon off the western coasts of Mexico, the U.S., and Canada—the northernmost point in this pinniped's migration. Individual sea lions have learned to follow fishing boats as they go to sea, and before a fisherman can reel in his line, they intercept his catch, leaving only the fish head or a bare hook.

Recently in Washington State, a score or so of wily males congregated at the bottom of a salmon run, gorging on the migrating fish and sparking protests from salmon advocates. Federal officials treated the most aggressive sea lions much like rogue bears, apprehending them and shipping them off to southern California. But the pinnipeds soon returned, and no easy solution to this conflict is in sight.

Female California sea lion, with pup

Northern Sea Lion *(Eumetopias jubatus)* The northern or Steller's sea lion is the largest otariid, with bulls ranging up to 13 feet long and weighing 2,400 pounds, while females are about 7 feet long and weigh 700 pounds. Propped back on their haunches with wide fore flippers, their eyes half closed and heads erect, their enormous chests bristling with guard hairs, bulls often appear resolute and serene. They should be nervous and apprehensive. Historically, Steller's sea lions have inhabited coastal areas of the North Pacific rim, with most of them occupying Alaska and the Aleutian Islands. In the last 30 years, however, their numbers have decreased alarmingly throughout most of that range. Since the mid-1980s, declines of up to 90 percent have been reported in at least three areas: the

Northern sea lions clash

western Gulf of Alaska, the eastern Aleutians, and Russia's Kuril Islands. Some rookeries, such as California's San Miguel Island, have been abandoned. Only southeastern Alaska's rookeries appear to be stable. In 1990 the National Marine Fisheries Service listed the species as threatened. A later analysis concluded that the Steller's would become extinct within 60 to 100 years if its downward trend was not reversed. In 1995 the species was proposed for endangered status. The reason for this decline is uncertain, but contributing factors include reduced prey (due to overfishing and climate change), shooting and accidental netting by fishermen, human disturbance of rookeries, and continued hunting by native peoples. The most likely cause is commercial overfishing of pollock, this species' principal prey.

Adult bull walrus struts his stuff

Family Odobenidae **Walrus**

Walrus *(Odobenus rosmarus)* Who does not know the walrus by its massive size, thick skin, stubby whiskers, and trademark tusks? Mature bulls can top 3,000 pounds. The genus name, *Odobenus*, means toothwalker—referring to this animal's reliance upon its tusks to haul itself out on land or ice. Occurring in both sexes, the tusks are stouter in males than in females. Like the incisors of rodents and elephants, they grow continuously. For at least 200 years they were thought to serve as digging devices for unearthing clams and other shellfish, the walrus's favorite food. But today, science sees that role as minor. Abrasions on tusks indicate that they are used not for digging or raking but are merely dragged along as the animals root with the upper edge of their snout for prey.

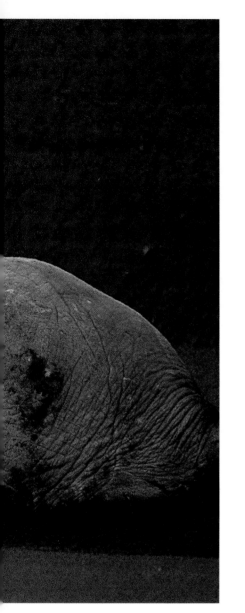

Walrus band resting but wary

The primary use of tusks, it seems, is in social communication. Like horns, antlers, large noses, and the like among other species of the animal world, tusks signal maturity, threat, and the probable outcome of a fight. Large-tusked males intimidate, while those less formidably endowed often move out of the way, deferring to the other's social dominance and avoiding the waste of a battle they know they cannot win. Tusks are important during the mating season and also are used to defend the young and even to attack hunters and their boats. They may be hooked over the ice edge like an anchor while sleeping in the water; they also serve as chopping tools for enlarging an ice hole or digging through soft or thin ice. The long-ago change from a largely fish diet to one based on mollusks may have played a role in the evolution of the upper canines into tusks.

Family Phocidae

True Seals

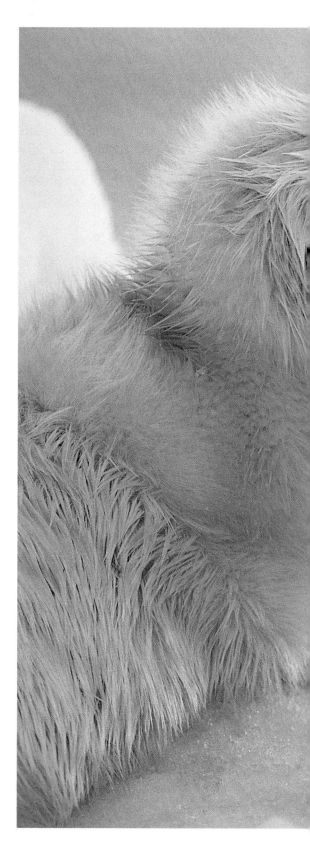

Nine out of ten of the world's 50 million or so pinnipeds are phocids, also called true or earless seals because they lack external ears. Since their hind flippers cannot be turned forward under the body, they pull themselves along the ground, using the fore flippers and lower body for traction. They swim by making sculling movements of the rear limbs. Much more diverse than otariids, they range in size from 4-ton, 20-foot-long southern elephant seals to 5-foot-long ringed seals, which weigh only 150 pounds.

Size differences between the sexes run the gamut from males being much larger than females (as with elephant seals), to moderately larger (hooded and harbor seals), to same size (harp and bearded seals), to actually smaller than females (monk seals). Similarly, members of this family can be highly gregarious (elephant seals), moderately so (harbors and harps) or solitary (ringed and ribbon seals). In general, the greater the size difference, the more sociable the species. Lactation and the mother-pup bond are brief, lasting no longer than six weeks.

Harp Seal *(Phoca groenlandica)* Long sought by humans for their delicate white coats, newborn harp seals have become a symbol for all exploited animals. Also called Greenland or saddleback seals, they were important in the settlement of Greenland and eastern Canada, providing many necessities such as food, fuel, and clothing. The annual harp "harvest" is the largest and most enduring seal hunt in the world.

Newborn harp seal pup, born on the ice, and its mother

Elephant seal pup takes in mom's rich milk

Northern Elephant Seal (*Mirounga angustirostris*) Nearly annihilated by 18th-century sealers seeking to render their oil, northern elephant seals have made a remarkable recovery. They were considered extinct until 1892, when a small herd was discovered on an oceanic island in Mexican waters. The herd grew, and by the 1930s elephant seals began breeding again on islands off the coast of Baja California and on the southern California mainland. Central California's islands were recolonized during the 1960s and 1970s, and the growth and recolonization of new rookeries continues today. The entire world population of northern elephant seals was estimated at 127,000 in 1991; all are descendants of the remnant Mexican herd. Can we assume that they have recovered fully from

Bull elephant seals battle for mating privileges

near-extinction, that the population is fit and viable and has been saved? It may not be so simple. Research shows that a species that goes through such a "population bottleneck" is changed. Genetic variability, every species' hedge against environmental uncertainties, is lost; the few survivors carry only some of the genes once present in the entire population. Today's northern elephant seals are so similar genetically that DNA fingerprinting cannot be used to determine paternity as it can in other seals, other animals, and humans. Our optimism must be guarded because the present and future populations of this majestic species—the largest North American seal—may be less adaptable than pre-exploitation ones. Bull northern elephant seals weigh up to 5,000 pounds; females tip the scales at a ton.

Sea Cows

Sea Cows

Order Sirenia

BY GALEN RATHBUN

During seven years as a research biologist in Florida, I spent hundreds of hours observing manatees in the spring-fed headwaters of Crystal River. I am still awed by how bizarre these animals are, and can't imagine anyone who has seen one not being struck by their odd, ponderous form and unusual behaviors. But the peculiarities of manatees go far deeper than a face only a mother could love.

There are only three living species of manatee, one in West Africa, another in the Amazon drainage, and the West Indian manatee—which occurs in northern South America, the Caribbean, and up the coast of Central America to the southeastern U.S. The closely related dugong is confined to the Indo-Pacific. These four make up the living members of the Order Sirenia, more popularly called sea cows. One measure of the distinctiveness of this group is that the closest living "cousins" consist of very different orders: elephants and hyraxes, possibly also elephantshrews and aardvarks. Even many biologists wonder at this odd assemblage.

The West Indian manatee has two subspecies, which have slightly different skull features and are found in different regions. The Antillean manatee occurs in northern South America, Central America, and throughout the Caribbean, while the Florida manatee lives in the southeastern United States. It is most common in Florida, where it inhabits rivers, bays, and seacoasts.

Sea cows are herbivores. They eat freshwater and marine plants, including seaweeds, algae, water weeds, and grasses. In fact, they will graze on almost any plant matter, living or dead. Florida manatees even eat mangrove leaves and acorns that have fallen into the water from overhanging trees.

Feeding on aquatic plants requires many specialized features not found in other marine mammals. Unlike whales and seals, which actively hunt their food and so must be strong, swift, or agile swimmers, manatees need to remain more or less stationary in the water column, in order to eat plants. As any snorkeler or scuba diver knows only too well, it is not easy to remain stationary in water. Currents, waves, and our own buoyancy move us about, often against our will.

How does an 11-foot-long, 2,200-pound manatee effortlessly "float" at whatever depth it wants? Bones help. Unlike most mammals, manatees do not have marrow cavities; their bones are solid and exceptionally dense, resembling ivory. Their mass offsets the buoyancy of insulating blubber and air-filled lungs. The lungs also play a role, for while the lungs of most terrestrial mammals are located in the chest or upper trunk, manatee lungs are long and thin, stretching nearly the entire length of the body cavity. This allows the animal to maintain, without any apparent effort, a stationary horizontal position in water.

Some plants that it eats, including many grasses, are very coarse. Some have large amounts of mud and sand attached to their roots and rhizomes. Chewing such grit causes teeth to wear out quickly—much sooner than the manatee's 50- or 60-year life span. But manatees, like elephants, simply grow new teeth as old ones wear out. Old molars are shed from the front of each tooth row, while new ones push up at the back and slowly move forward along the jaw, as though they

previous pages: **Adult manatees socialize with a youngster**

are on a slow conveyor belt, moving forward about one-third of an inch a year, until each drops off the end. Elephants have only three molars per tooth row; once the third wears out, they can no longer chew their food. Manatees, however, continue to grow new molars throughout their lives.

Herbivorousness also influences other aspects of manatee biology, such as habitat. Almost all plants require sunlight, but light does not penetrate water very well. This means that aquatic plants—and the manatees that eat them—are confined to shallow waters. Although manatees can swim far out to sea, they rarely do because the plants they seek don't grow in ocean depths.

Compared to fruit, grain, and meat, most aquatic plants are not highly nutritious. Florida manatees must spend six to eight hours a day feeding, just to gather enough nutrients and energy to survive. This relatively low-energy diet helps explain why manatees are warm-water creatures, normally not venturing into waters below about 68°F. In the U. S., especially during winter, this means Florida.

Will Florida's manatees endure? Fortunately we do not compete directly with them for food, nor do they pose a threat to us. (If they had, we probably would have eliminated them long ago—like the wolves, bears, and mountain lions that have vanished from so many parts of North America.) But manatees occupy the same waters that more and more people use for boating, cooling power plants, and disposing of waste. They normally range three to ten feet below the surface. Each year, Florida boaters accidentally kill dozens of them. In fact, most manatees carry wounds or scars inflicted by collisions with boat hulls and propellers. Ironically, there is widespread support for these animals among Floridians. But manatees haven't yet learned to avoid people—and they remain tied to Florida's warm, shallow waters as strongly as we seem to be.

Like us, manatees reproduce slowly, giving birth to a single offspring every two or three years. This low rate and the numerous accidental deaths due to humans has moved the U. S. Fish and Wildlife Service to list the animal as endangered. A 1997 census tallied over 2,600 manatees in Florida—but it's unclear whether this is an increase or decrease. One thing is certain: as long as the area's human population continues to swell and to develop the coast, manatee habitat will decline.

Just look at Crystal River; when my manatee studies ended there in 1985, the waters were so clear that the pure white sand bottom seemed close enough to reach from the boat—though it lay ten feet down. Twelve years later, there were still plenty of manatees (and many more people). But the water was so murky that I had trouble seeing the bottom at all. The sand was covered with dark brown muck, and most of the aquatic plants were gone. Alas, the crystalline headwaters of Crystal River seemed to be turning into a mud hole.

Nobody knows how severely these changes will impact the 300 or so manatees that use Crystal River during the winter; only time will tell. In the meantime, there is little doubt that we are losing the unique conditions that once allowed all of us to observe, photograph, enjoy, and better understand these unusual mammals. ■

Family Trichechidae Manatees

Manatee (*Trichechus manatus*) In the crystal clear waters of a Florida spring (above), a mother and calf manatee demonstrate their playful curiosity as they "mouth" a boat anchor line. Unfortunately, such natural curiosity can get them into trouble. Several times a year rescuers from Florida oceanaria are called out to capture entangled manatees and remove buoy or anchor lines from their flippers.

A resting manatee (opposite) slowly rises to the surface for a breath of air, after pushing gently off the bottom with its flippers. Manatees are not always loners; at warm-water sites during the winter, they often aggregate into temporary groups that intensively nuzzle and rub each other (pages 146-147), although interactions of this sort are usually short-lived. In warmer weather, manatees are normally widely dispersed in murky waters where they are difficult to see. Their best defined social unit is that of a female and her calf. Small calves rarely stray far from their mothers, nursing several times a day from a single mammary gland located at the base of each flipper. Normally they are weaned at about two years of age.

Other social behavior exists but is more ephemeral. For example, females that are in heat attract up to a dozen adult male consorts during their two-week-long estrous period. The males that make up these "mating herds" change constantly over that time, pushing and shoving each other as they try to mate with the female. Smaller, younger males also form herds that engage in shoving matches, but only for a few hours at a time. No doubt such play-fighting helps them prepare for the day when they will compete in earnest with other adult bulls for females.

Algae-covered manatee in a clear spring

Ungulates

Ungulates

Order Artiodactyla

BY VALERIUS GEIST

Giants still walk among us. Take almost any large mammal that Eurasia shares with North America, and you will find it larger in the New World. The Siberian moose is big, but the Alaskan moose is bigger. The bison is the largest mammal in Europe, but it grows even larger in North America. While both the North American elk, or wapiti, and mountain goat have many close relatives in the Old World and none on this continent, each is the largest of its group.

Why are North America's mammals so much bigger? Perhaps it's because they originated in Eurasia, entering the New World via the Bering land bridge that intermittently connected Alaska and Siberia during the Ice Ages. Cold climates encourage large size, because a big animal has less skin surface in relation to its body volume than a small animal, and therefore retains heat better. But also, the New World was full of many large predatory species; prey animals that evolved into larger, faster, and more powerful forms were more likely to survive.

Although this size pattern extends to other orders as well, it is most apparent in the Artiodactyla, the even-toed ungulates. Members of this group, despite their formidable size, are characterized by a vegetarian diet and a readiness to run from danger. They rarely stand and fight off predators—though, as we shall see, much of their life centers on confrontation with one another.

The stress of running led to the development of defining characteristics in the feet of Artiodactyla. Of the five original digits shared by most mammals—indeed, by many vertebrates—the first has totally disappeared; the second and fifth also are absent or have become greatly reduced, as in the form of dew claws. The main axis of each foot passes between the third and fourth digits, thereby producing the familiar cloven hoof. (In contrast, the horse—a member of the odd-toed order called Perissodactyla—carries its weight on a single toe.) Artiodactyls also possess another peculiarity: The ankle joint of the hind foot consists of a unique hinge mechanism that ensures remarkably free movement and flexibility.

This order comprises three living suborders. One, which includes camels and llamas, no longer occurs in North America. A second suborder is represented on this continent only by two species of peccary. All other North American artiodactyls—deer, pronghorn, bison, mountain goat, musk ox, bighorn sheep and more—belong to the ruminant suborder. Ruminants have a more sophisticated digestive system than peccaries, one that can exploit plant cellulose for energy. No mammal produces enzymes that digest cellulose; ruminants, however, culture microorganisms—both bacteria and protozoa—to do it for them. The microbes grow within the rumen, a part of the gut, in which fodder ferments by bacterial action. Fermentation produces fatty acids and other products that can enter the animal's bloodstream directly from the rumen. Bacteria and residue from the rumen later pass into other parts of the gut and are digested as well, further supplementing the host animal's protein, mineral, and vitamin requirements.

Most true ruminants have projections on their heads. Those of the bovid family carry tough horns covering a bony core; these are permanent, increasing in length

previous pages: **Massed shoulder to shoulder, musk oxen present a formidable defense**

and size each year. Examples are mountain goats and sheep, musk oxen, and bison. Members of the deer, or cervid, family rely on antlers—bony structures grown and shed annually. The pronghorn, a species unique to North America, has horns with a permanent bony core but a keratinous sheath that is shed each year. In horned species, both sexes usually carry them, though generally they are much less massive in females. Antlers normally do not develop in female deer, elk, and moose, but female caribou do possess a smaller version of the male form.

Horns and antlers serve both as weapons and for display purposes. They may symbolize prowess and ability; well-developed ones show that the owner is a good forager, one that can obtain more food than the minimum needed to survive. Males with the biggest projections achieve dominance and breed with the most females; their traits therefore are passed on to more offspring. However, my own studies with mountain sheep have shown that rams with the largest horns—that is, the most dominant, the most virile, and the most successful individuals—have shorter lives than other males. They simply exhaust themselves fighting and mating. Less active and less prosperous males have longer life expectancies.

If antlers are so important, then why shed them? During the mating season, bucks busy at playful sparring, serious fighting, and courting have little time to forage. By the end of the rut they are exhausted, weakened, and vulnerable. Their antlers help them stand out from females; perhaps New World deer evolved the trait of shedding such standout racks in order to blend with the crowd.

Antlered artiodactyls have other defense strategies as well. The white-tailed deer favors dense brush; if discoverd, it erupts from cover, then dashes off, trying to get so far ahead that its scent may evaporate and its trail dim by the time predators catch up. It often runs through water or swampy spots to avoid leaving its scent. It also likes to run where other deer may be lurking—and let someone else get chased for a while. Such habits make for a nervous, fidgety species.

Mule deer, however, are more likely to head for the brush, coolly trying to stay hidden. If flushed, they "stot"—jump straight up—and may then head off in different directions. When a predator is close at hand, however, they must not jump until the last possible moment; otherwise the predator can redirect itself and might make a successful kill. While whitetails prefer to rush downhill, using gravity to quicken their speed, mule deer often bound uphill. They gain elevation fairly easily, while the predator is forced to clamber up at far greater energy cost. Mule deer also can leap over a boulder or bush that the pursuer must go around. Such escape tactics call for coolness in timing—and make for a calm, collected type.

On the tundra, on open plains, or in other places with scant cover, a lone newborn artiodactyl can be particularly vulnerable to predators. But if many young are in one area, their numbers serve to swamp the enemy; a few calves succumb to predation, while most survive. Thus, genetic selection among herding animals such as reindeer and tundra caribou has resulted in females that not only reach breeding state all at the same time, but also give birth almost en masse.

But among species whose young instinctively hide, such as white-tailed deer, females give birth at different times. This reduces the risk that predators, taking one youngster, will learn to expect to find others. Thus, the birthing season of such "hiders" tends to extend beyond that of herders. So does the rutting season. Warm climates, with their milder conditions for vulnerable newborns and longer periods of nourishing vegetation, present more opportunities for rutting and giving birth. In such regions, females readily extend the birthing season even more.

Except for the pronghorn, single young are the rule among New World animals that live in herds and run for their lives. That's because the larger the offspring at birth, the more likely it will be able to outrun predators. Also, it will grow out of its dangerous juvenile period faster if it does not have to compete with a sibling for its mother's milk. The baby wildebeest, or gnu, of Africa is a classic example; five or ten minutes after birth, it can keep up with a herd that runs 25 miles an hour.

While the ungulate mother bears and cares for her young, the father contributes only his genes. Therefore, this contribution should come from a superior male. To be fit, one must maintain access to resources, particularly scarce ones. Fighting over food can be costly in time and energy, to say nothing of the danger from horn or antler. The cheapest way to maintain access to needed resources is to establish dominance over competing individuals. The most dominant animal gets the biggest share, while other animals get progressively less.

A male ungulate's aim, of course, is to ensure access to females. Some species, such as mule deer and moose, are sequentially polygamous: the male stays with one female until she is bred, then leaves her for another. In contrast, an elk buck herds females together into a harem with which only he mates.

The bull elk advertises. He bugles, trying to outbugle his competition, and if one advertises nearby, he will try to shut him up or chase him off. The bigger the elk's body, the more resonant his call. Consequently, females generally cluster to the bull with the deepest voice—and the one that other bulls can't shut up.

Ungulates fighting for dominance can cause each other serious wounds and even death, however. The sharp horns of mountain goats, for example, jab and puncture, and so the males have evolved an incredibly tough shield—hide an inch thick—on their rumps, where the most blows land in their side-to-side, head-to-haunch type of fighting. Even so, deep wounds can result.

Elk charge with sharp polished antlers; punctures and broken necks may occur. Bighorns clash head-on. They have evolved multi-roofed skulls, thick facial skin, and incredibly sturdy neck tendons to withstand tremendous impacts. Moose can kick from both ends, an effective defense against wolves and other predators. One report relates that a bull moose, using its hind legs, struck a man standing on a corral chute eight feet above the ground!

During male-to-male jousts, injury to the victor can be as severe as it is to the vanquished. Real battles, therefore, are relatively few. Instead, ungulates have adopted more subtle and less costly methods to establish superiority. One is called dominance displays. Essentially, such displays involve an animal showing off the size of its body or its weapons without making a move to use them. The next higher level consists of threat-making, that is, bringing the weapons into readiness. Deer that might normally rise on their hind legs to flail a rival with their forelegs may first threaten them by lifting the head or fore body, by raising a leg, or by stamping the ground. A cow elk, which attacks by biting, will pull back her lips and grind her teeth. A dominant bighorn ram might threaten an inferior by jerking his horns downward in order to frighten off the competition. But to a serious rival, he will display his horns with his head drawn up and back, causing his neck muscles to bulge as he advertises his formidable weaponry.

For most large mammals, displays consist largely of maximizing the apparent mass of their bodies. They usually do this by turning broadside and erecting specialized hairs, such as the hair of the mane or along the spine. Their stance also calls attention to distinctive markings and makes the animals appear as large and conspicuous as possible.

Male red deer, European relatives of elk, enhance the size of their antlers by horning shrubbery and collecting vegetation on their racks, which makes them look even larger than they are. During the rut, a bull elk urinates on his underbody, then wallows in wet soil. The mud darkens the animal along his entire length, including the mane, making him visually more impressive.

Such displays alone, however, aren't always enough to ensure success. Dominant animals must also have credibility: Now and then, they must back up their displays and threats with real force. Among artiodactyls, the ultimate credibility stems from contests of strength. It pays the dominant animal to periodically reinforce his position—just as it pays subordinate animals to test the dominant's competence from time to time. So it is that sparring and even serious fighting often occur between displaying artiodactyls.

Interestingly, mule deer form social bonds as a result of their playful sparring matches. A superior male may engage a lesser one in successive bouts that can last for hours—much the same as in other species. But if the inferior mule deer gets harassed by other lesser males, then the superior comes to his rescue.

Normally, sparring and serious fights end whenever the subordinate breaks off the action and exhibits some submissive act. Pretending—with exaggerated motions—to graze is an almost universal signal among hoofed animals that says, "Look, I'm really peaceful; I'm no threat to you." Another common sign of submission: acting like a female.

For ungulates, the cumulative effect of their weapons, their ability to learn, and the preference of individuals to live in a social milieu where roles are understood and played out—all this combines to create a dominance hierarchy even we can easily recognize. Artiodactyls sometimes act just like us. ■

Collared peccary mother and child

Family Tayassuidae Peccaries

Also known as javelinas (due to their spear-sharp tusks), peccaries superficially resemble pigs but belong to a separate family, one which occurs only in the Western Hemisphere. Unlike pigs, they have fewer teeth (38 rather than 44), partially fused foot bones (for instant acceleration), and a more complex stomach (which enables them to digest fibrous foods such as cactus, spines and all). The family consists of three living species, which range from the southwestern U.S. to central South America.

Collared Peccary (*Pecari tajacu*) Three dark look-alikes drink from the same water hole (opposite). Which is male, which female? Differences between the sexes are minimal in this species; both possess short, sharp tusks and are similar in size and appearance. Dark colors commonly occur among large herbivores that tend to stand up to predators or to confront one another in territorial disputes. Collared peccaries occur from Arizona and Texas into northern Argentina. Adults weigh 30 to 60 pounds and, like many other large mammals that defend their living space against those of their own species, have evolved weapons that can inflict severe wounds on opponents. Peccaries also live in organized and fairly permanent groups of 5 to 15 members, largely for security reasons. A musk gland in the rump yields a strong odor that marks the home range of the herd and helps members identify each other. They often rub against each other.

Collared peccaries share a water hole in south Texas

Family Cervidae **Deer**

Long-legged and graceful, sometimes delicate and sometimes formidable, deer are perhaps best known for their antlers, hornlike outgrowths of true bone that are carried principally by the males and are shed and regrown annually. Worldwide, the family includes some 17 genera and at least 45 species, with additional ones being discovered in the tropics. North American members vary from tiny Key deer of the Florida Keys to huge Alaskan moose.

Cervids—deer—are opportunists, surging to prominence during the last two million years. Many thrive on disturbed, nutrient-rich ecosystems; they benefited from the ecological turmoil caused by great climatic changes, by glacial actions that enhanced the fertility of soils, and by extinctions of certain large herbivores and carnivores. Originating in tropical Asia, this family rapidly evolved to occupy cold environments, including high Arctic and alpine areas. Despite its diversity, however, it remains tied to forests, flood plains, and savannas, rarely intruding into steppes and deserts, which are the the chief domain of bovids, horses, and camels.

Males of all North American species, as well as both sexes of caribou, carry antlers, which grow rapidly in summer. At first they are soft, well supplied with blood, and covered in thin skin with fine hairs that make them seem draped in velvet. As growth stops, circulation of blood to the antlers ceases; the skin dries and is rubbed off. Later—usually in winter—the bone at the base of each antler breaks down, allowing the antlers to drop. Antler growth depends on an animal's health, not its age; a large rack advertises the owner's ability to find the best foods and divert them into antler growth. Such qualities may be attractive to females, which produce offspring that must be well developed and quickly achieve running speed.

Elk or Wapiti (*Cervus elaphus*) Called wapiti by Shawnee Indians, the elk is a subspecies of red deer once widespread in Eurasian forests. It is, however, more adapted to life in the open country and is capable of swift and enduring runs. It has a larger body—up to 1,100 pounds—a more ornate coat, more advanced antlers, and larger teeth than its Old World relatives. It came from Siberia to Alaska across the Bering Land Bridge during the Ice Age, entering southern North America only at the end of the last glaciation, after most large mammals native to that region had become extinct. With no competitors or predators to deter it, and being well equipped to cross rivers and mountains, elk occupied habitats as diverse as Pacific rain forests, eastern woodlands, sagebrush plains, prairies, and alpine parklands. Historically found from southern Canada to Mexico and all across the lower 48 states, elk were nearly wiped out by market hunters in the 19th century. Subsequent protection has led to locally abundant herds, mainly in the mountains of the West, where aggregations of up to a thousand individuals may occur in limited winter habitat. Born in late spring, the calves live with their mothers for several weeks before joining herds led by older cows. Adult bulls live alone or in small groups throughout the summer. By September, antler growth is complete and they bugle and joust with one another to attract harems of up to 20 females.

Bull elk in velvet roams Yellowstone

Moose *(Alces alces)* Antlers lowered, bull moose hurl their heavy (up to 1,800-pound) bodies at each other in a violent though rare fight. Their many-tined antlers serve as protection, catching and holding those of the opponent. They also lock the heads together, allowing for powerful wrestling. Tines may splinter; so may entire antlers, leaving the contestant unprotected. Even with antlers intact, fighting leads to injuries—despite a thick hide over vulnerable areas. During a single mating season a bull may accumulate 50 or more antler punctures, which heal only at great physiological cost. Some animals die outright; at times opponents lock antlers permanently, and both succumb. Far more commonly, however, bulls

previous pages: **Cow moose grazes the shallows of Alaska's Wonder Lake**

Bull moose clash during the annual fall rut

of very different size and rank take on each other in harmless, friendly tussles.

This species entered North America from eastern Siberia about 10,000 years ago, just before the Bering Land Bridge flooded, and spread across Canada and the northern U.S. It leads a largely solitary life in the woodlands, browsing on twigs, leaves, bark, and shrubs. Its peculiarly shaped muzzle apparently allows it to feed efficiently on submerged aquatic vegetation. The hairy and pendulous "bell"—a skin flap under the chin—is soaked in urine during the rut and serves as a scent dispenser. Males compete for females one at a time during the fall rut. The new generation, usually single calves but often twins or even triplets, arrives in spring.

Resting mule deer

Mule Deer *(Odocoileus hemionus)* Resting in cover, its large ears gently twisting and turning, this mule deer buck scans the landscape for sounds of distant danger. Should it detect harm coming its way, it will rise calmly and walk leisurely to safety. It also may stot—bound stiff-legged into the air with all four feet off the ground—as it retreats. This unflappable deer quickly learns to distinguish what is harmless from what is not and adjusts accordingly. Where not hunted it becomes remarkably tame, readily accepting suburban landscapes as habitat. It feeds on a great variety of plants, and occurs in the millions from northwestern Canada to northern Mexico, usually living alone or in small groups. Males weigh up to 400 pounds; females are much smaller. Some researchers believe that mule deer may have originated as hybrids from female white-tailed and male black-tailed deer, possibly as recently as the beginning of our current interglacial period.

Whitetails, alert but not yet alarmed

White-Tailed Deer (*Odocoileus virginianus*) A great biological success, this cervid ranges from Canada to well below the Equator, in South America. No other artiodactyl can match its huge north-to-south distribution. It is a roving opportunist capable of quickly exploiting disturbed ecosystems, where plant successions are in early stages. It thrives in river floodplains and deltas, in areas that have been logged or burned, and in agricultural, suburban, and even urbanized regions. It does require some dense cover where it can hide by day; it is most active around dawn and dusk.

The whitetail's dietary habits resemble those of mule deer, though it is more of a browser and tends to be smaller, with adult males weighing up to 300 pounds. A diminutive subspecies in the Florida Keys—the so-called Key deer—averages just 80 pounds. It has been endangered by habitat loss and increased automobile traffic.

Bull caribou in rut

Caribou *(Rangifer tarandus)* Known in the Old World as reindeer, the caribou is the greatest migrant and most highly evolved runner of the deer family. It thrives mainly on the tundra from Alaska to Greenland, but also occurs in forests as far south as Idaho. Some herds migrate annually over 600 miles from tundra to timber, eating mostly lichens and congregating in groups of up to 250,000 individuals. To run on soft snow or spongy ground, the caribou has evolved its own "snowshoes" in the form of greatly expanded hooves. In the water these serve as paddles, making this animal a superlative swimmer—a skill especially important in summer, when snowmelt covers the tundra with wide, shallow lakes and swift rivers.

Moving efficiently is critical in escaping its primary predator, the gray wolf, also broad of foot and a relentless traveler. The perpetual contest between hunter and hunted has led to enormous antlers in this cervid, the only deer in which both sexes wear a rack. Males use their antlers and a decorative hair coat (above) to advertise their competence in acquiring the resources necessary for such growths. Females give birth to calves that grow quickly and can soon outdistance wolves.

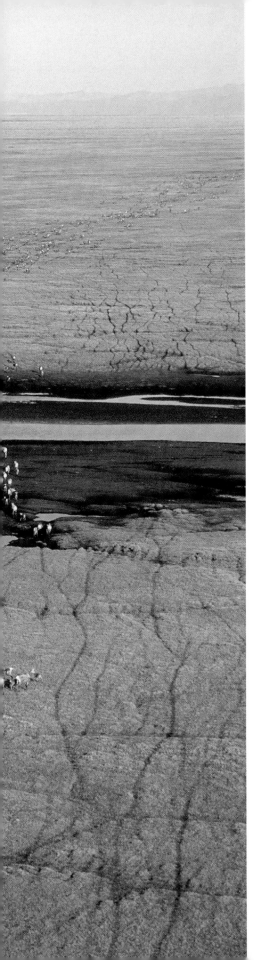

Caribou mass on Alaska's North Slope

Female and male pronghorn, ready to bolt

Pronghorn

Pronghorn *(Antilocapra americana)* Equipped with eyes larger than an elephant's, and with legs that can outrun any other North American mammal, the pronghorn is superbly designed to escape predators. It can achieve speeds of over 50 miles an hour, and it cruises at 25 to 30. It is not merely fast, but fast over obstacle-strewn terrain, which demands precise leg placement. Thus pronghorn eyes are set wide apart, to accurately judge distance, and the head is held high while running, to scan the landscape ahead. The acute dangers faced by pronghorns also fostered intelligence in this species, as well as a high rate of reproduction and early maturation. While most of North America's megafauna died out during the Ice Age, the pronghorn survived.

This is the last of an ancient New World ruminant family characterized by forked horn sheaths, which grow from permanent horn cores in both sexes and are shed annually. During breeding season in late summer and fall, adult males compete for control of groups of females. The young—usually twins—are born in spring and grow rapidly; at only four or five days of age, they can outrun a person. Adults weigh 80 to 150 pounds and reach a shoulder height of about three feet. The white hairs of the rump patch can be raised as a warning signal visible miles away. Although usually able to escape predators, the pronghorn was nearly wiped out by human hunters in the early 1900s. Protective laws have since allowed it to recover in some areas; perhaps a million exist today.

Family Bovidae **Bovids**

With some 47 genera and at least 138 living species, the bovids are globally the most successful family of artiodactyls. They boast a huge geographic distribution and a great diversity of ecological adaptations. Also called "hollow-horned" ruminants, they grow permanent horn sheaths over bony horn cores that project from the front of the skull. These horns are borne by both sexes in most species, though they often are more developed in males, especially sheep. They assume many shapes, from short daggers and slashing sabers to wide-spreading arches, knotted lyres, and twisted spears. Horns function not only as weapons, but also as shields to catch and parry attacks by a rival, as grappling hooks, as locks to bind heads together in wrestling matches, and as showy luxuries to impress females.

Bovids have an edge on cervids when exploiting relatively dry, nutrient-poor grasslands, but even they are not very successful in deserts. They are ruminants, able to regurgitate food for a second chewing. Those that feed on grass pull it rather than bite it off, which enables them to get at the tender lower stems.

This family originated in the Old World; its evolutionary diversity remains greatest in tropical and subtropical Africa and Asia. Bovids entered our continent only with the beginning of the Ice Age, and never reached South America on their own. Despite evolving several large-bodied species, all highly specialized to deal with North America's many large predators, bovids have played a subordinate role to other native herbivores such as the early horses, camels, and ground sloths. The number of North American bovid species, five, is comparatively low; most of those have been greatly reduced by human hunting and habitat dirsruption. Even so, bovids include our most abundant domestic mammals: cattle, goats, and sheep.

B i s o n *(Bison bison)* Weighing up to a ton, this is our continent's largest living land mammal. Its ancestors originated in Eurasia, crossing into North America as early as 500,000 years ago. An estimated 50 million bison roamed North America in historic times, ranging from northwestern Canada to Florida.

The bison concentrates its bulk in its forequarters; the heavily muscled neck supports a massive, low-slung head. Primarily a grazer, it has been most successful in the Great Plains. Although known for vast congregations, the basic social unit is a small band of females and young. During mating season, adult bulls join these bands and fight fiercely for mates, mainly by ramming each other head-to-head, bellowing deeply, and hooking their horns. Serious wounds, even death may result, but usually one bull turns aside in submission. Victorious males tend one female at a time, for several days each. Single calves are born in the spring.

To subdue some Native American peoples who subsisted largely on bison, and to make room for agriculture, bison were systematically destroyed during the 19th century. By 1890 fewer than 1,000 were left, in two remnant herds. Subsequent protective measures fueled a remarkable recovery; some 200,000 individuals range parts of North America today. Most are in a captive or semicaptive state, although a number of populations exist in natural ecological contexts.

Bison struggle through Yellowstone's deep snows

Desert bighorn ram and lamb

Bighorn Sheep *(Ovis canadensis)* The trademark horns of this species become huge only if the ram finds sufficient high-quality food to grow a sizeable horn segment each summer—and only if he manages to scale steep cliffs consistently over many years, to escape predators. Large horns thus are a measure of success and are displayed conspicuously to females during courtship in the fall and early winter. Males expert at finding the best food safely make the ideal mates, as the lambs they sire are more likely to be spunky and vigorous, quick to hop on sturdy legs and bounce up cliffs—and occasionally on other bighorns (above).

Like a deer's antlers, a ram's horns also are crucial in confrontations over ewes. A large set may weigh 30 pounds, representing more than 10 percent of the entire

Bighorn line out in the Canadian Rockies

animal's mass. Horns are used like sledge hammers, the opponents rearing up and then crashing together head-on. A double-layered skull and shock-absorbing connective tissue help them survive repeated blasts. But the bigger the hammer, the harder the blows—and the more likely the male will triumph in battle, or even avoid a fight merely by showing his size. A ram with trophy horns also attracts young males, which follow him because his horns signal that he knows of places with superior food or security. The youngsters gain the ram's acceptance by acting much like females in heat, for no ram would strike a receptive female. Over a million bighorns once roamed from western Canada to Mexico; barely 5 percent of that remain, due primarily to transmission of diseases from domestic sheep.

Point of impact

D a l l ' s   S h e e p *(Ovis dalli)* Here, the attacking ram is on the left, the defender on the right. Fights usually are initiated by the smaller, more insecure contestant, the one most likely to lose. He takes advantage of the terrain, however, starting his bipedal run from uphill and hurling himself at the defender while swinging his horns sideways so as to strike with the edge of one horn. It's sledge hammer and karate blow all in one. But the defender rises nimbly to catch the blow squarely between his horns and against his armored skull.

Much of the impact is absorbed through a complex hinge mechanism linking skull and neck with a massive tendon. The remaining force is dissipated as the defender allows himself to be propelled backward. Each combatant then displays his horns, showing the other what delivered the blow that was felt. Consequently, rams soon become accurate judges of horn size, and dominance fights rarely occur between those that are not equally endowed. Contests end when one ram stumbles after a blow, rather than alighting gracefully. He is the loser, but he will not be expelled from the band, although the winner will treat him as if he were a female.

Such contests may break out at any time and can last for hours, but are most common in the autumn mating season. Adult rams weigh 165 to 250 pounds, less than their bighorn cousins. Their horns tend to spread and spiral more, the tips pointing outward from the face. After mating season, males roam in small bands from one grazing patch to another.

Two rams posture before the clash

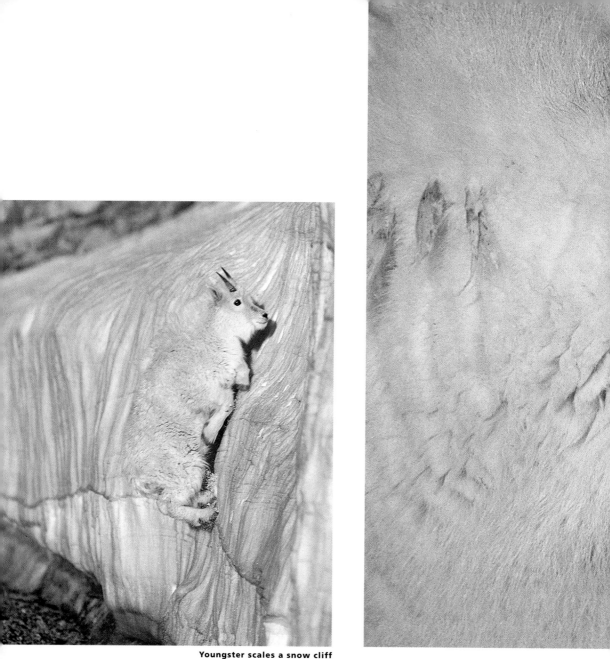

Youngster scales a snow cliff

Mountain Goat *(Oreamnos americanus)* While mountain sheep are skilled jumpers, mountain goats are methodical climbers—an appropriate skill among cliffs where the footing may be hidden by snow for more than eight months of the year. Large hoofs enable them to walk or crawl over crusted snow. Goats rarely leave the vicinity of cliffs, scurrying back at the slightest hint of danger.

Immediately after the mating season, in early December, females with kids claim the best cliff habitat and expel all other goats, males included. Although adult males, at 190 to 280 pounds, are a third larger than females, they remain submissive to females throughout the year, because resistance can jeopardize their investment in reproduction. When courting they approach on their bellies, softly squeaking like kids, ready to crawl away if threatened. Younger males, not having bred, may

previous pages: **Mountain goat mother and kid on the precipice**

Among its own kind

resist females, which are more aggressive towards them. Goats have fragile skulls; they do not butt heads but instead stab with daggerlike horns, often aiming for the belly, groin, haunches, or chest. They avoid injury by evading opponents' thrusts and by absorbing them with a thick, tough hide, which may exceed an inch thick in places. Due to the potential for severe injury, goats fight only exceptionally. Even play can be dangerous, and females are extremely watchful, protecting their young in altercations with other goats. Mothers give birth in late spring, usually to a single kid, and remain with it until they again give birth. Due to their inaccessible habitat, mountain goats have been less affected by humans than have other large mammals of North America; they still occur throughout most of their original range, from Alaska to northern Oregon and Montana, and have been introduced to Colorado.

Musk ox *(Ovibos moschatus)* Following an instinct that protects them against wolves but leaves them vulnerable to humans, musk oxen form a defensive ring in the face of danger. Adults block access to the vulnerable calves and may suddenly charge, hooking at wolves with curved horns. Alone, even a large bull may succumb to a persistent and agile wolf. Like caribou, musk oxen have adapted to the severe tundra of the high Arctic; unlike caribou, they prefer sedges, grasses, and willows to lichens. Short, stout legs and large hoofs make musk oxen excellent climbers in cliffs and glacial rubble, and they are surprisingly nimble for their size. Adult bulls, which weigh 750 to 800 pounds, compete for control of female bands during the

Musk oxen circle up on edge of the Bering Sea

summer rut, repeatedly charging each other at up to 25 miles an hour and banging their armored heads together in spectacular clashes. This ancient species, Eurasian in origin, remains from the earliest entry of bovids into North America nearly two million years ago. It once occurred from northern Alaska through Canada to Greenland. But it is slow to mature and reproduce, bearing one calf every one or two years, and is no match for humans, who by 1900 had extirpated it from Alaska and nearly from mainland Canada. Subsequent protective measures enabled a slow recovery; there now are over 100,000 musk oxen in Canada and about 15,000 in Greenland. A small population has been successfully reintroduced in Alaska.

Aliens

Aliens
Introduced Mammals

BY RONALD M. NOWAK

Outsiders that do not belong. Strange beings from far away. Destructive invaders. Such are the thoughts invoked by the above title. But most of North America's alien mammals are familiar, even likable creatures deliberately brought here from somewhere else, by people. That such introductions often have gone awry is not the fault of the animals involved.

It often is difficult to judge what does not belong. Many mammals we consider native to North America—brown bears, elk, bighorn sheep, and others—are in fact relative newcomers that reached this continent only a few tens of thousands of years ago, by crossing the Bering Strait after sea levels fell. In contrast, horses evolved here over tens of millions of years, flourishing until they were killed off, probably by human hunters who arrived late in the Ice Age, no later than 12,000 years ago. When brought to North America by Europeans only 500 years ago, modern horses quickly reverted to the wild state and occupied vast regions in relatively little time. Should they really be considered aliens?

What of those animals native to parts of North America that later expanded their ranges through human introduction or habitat modification? They include the coyote, armadillo, northern pygmy mouse, and collared peccary. In their newly occupied areas, are they any less alien than those coming from other continents?

A 1997 Texas Tech University checklist of the mammals of the continental United States and Canada includes 28 species "not native to North America" but that currently "occur in the wild state in numbers sufficient to justify listing them." Those species, which are included in the comprehensive species list that begins on page 192 of this book, fall into four broad categories.

First are the commensals—the house mouse, black rat, and Norway rat. All normally live in close association with people, who unwittingly brought them here aboard ships as long ago as colonial times, and who continue to provide them food and shelter. Of all introduced species, these three best fit the image of unwanted, destructive aliens. They consume enormous quantities of food stored for people and livestock, and contaminate even more. They damage human habitations as well as native wildlife habitat. In the U.S. alone, direct economic losses to rats and mice run to over one billion dollars annually. These species also spread many diseases— bubonic plague, typhus, leptospirosis, tularemia, and others—which have taken more human lives over the past ten centuries than all the wars and revolutions ever fought. Bubonic plague, transmitted by rat fleas, killed a third of Europe's population between 1347 and 1352. More recent outbreaks occurred in San Francisco (1902 to 1941), Galveston (1920 to 1922), and New Orleans (1912 to 1926).

A second alien category comprises domestic mammals that have established wild-living, or feral, populations. Longhorn cattle, for example, once roamed parts of the West before disappearing in the wake of settlement. Other feral cattle still persist in scattered areas of the U.S. Great herds of wild horses—mustangs—were nearly wiped out earlier in this century by commercial hunters who shot them for hides and for use as dog and chicken food. Their estimated numbers fell from a

previous pages: **Enduring symbol of the Old West, modern horses are an Old World introduction**

million in 1925 to just 17,000 by 1971. Since then, federal protection has allowed modest recovery of mustangs, but not without controversy. Some people regard them as interlopers that overgraze livestock and native wildlife ranges; others consider them living history worthy of protection. The same is said of feral burros, which occupied the Southwest following their initial import and use as pack animals. As recently as 1994, some 6,000 burros and more than 36,000 horses ran wild in the U.S.

Far higher numbers of feral pigs—perhaps 500,000 to 2,000,000—inhabit our country, ranging from Texas to the Carolinas as well as throughout California. Although sport hunters take over 100,000 feral pigs annually, the animals continue to thrive, at our expense. They are detrimental to agriculture, forestry, and native wildlife. The same is true of feral goats, also widespread in the United States.

The oldest known domestic animal is the dog, a direct descendant of the wolf. New DNA studies suggest that dogs originated perhaps 130,000 years ago; most authorities believe that humans first domesticated them only about 14,000 years ago, in southern Asia. There, to this day, primitive pariah dogs still lead a semi-domestic or feral existence. Early human seafarers likely took such dogs to regions such as Australia, where they gave rise to dogs we call dingoes. Biologist Lehr Brisbin, in South Carolina, believes similar early canines today are represented in this country by a feral population he discovered along the Savannah River.

While the dogs of the original Americans have disappeared, some fully feral packs descended from European breeds continue to occur sporadically in parts of the U.S. and Mexico, patrolling regular home ranges, digging dens, and producing young. They feed on carrion, small animals, and sometimes deer, but are not a serious threat to wildlife populations or to people. Nearly all of the up to three million attacks reported annually in the United States—and the twenty or so human deaths—are caused by uncontrolled captive but fully domestic dogs.

Rabid dogs are traditionally seen as an archtypal menace, but in the U.S. other animals—particularly house cats—have recently been more dangerous carriers of rabies. Feral cats also are more populous than dogs and are a substantially greater problem for wildlife, especially birds. Cats were domesticated much more recently than dogs, probably no earlier than 4,000 years ago in Egypt, and they seem to revert far more readily to a wild existence.

An interesting though ecologically insignificant animal introduction to North America was the camel. Touted as efficient desert transportation, 34 camels were brought to the Southwest by the U.S. Army in 1856. Others were imported later, some by civilians who predicted that they eventually would supplant beef cattle. Rail construction helped burst that bubble, and some camels were released. They and their descendants wandered America's deserts until at least 1905; one recaptured individual lived in a zoo until 1934.

The third and largest category of aliens comprises wild mammals introduced for sport hunting, usually without thought of other potential consequences. One,

Nutria youngsters laze on a Louisiana log

the European rabbit, is the same species that has ravaged the vegetation of England, Australia, and many oceanic islands, to the detriment of countless native animals. A particular population, established about 1900 on the San Juan Islands north of Puget Sound, burrowed so extensively that the bluffs of one island crumbled into the sea. Luckily, later introductions to the Midwest seem to have failed, although the larger European hare has succeeded in some areas, becoming a pest to farmers in Ontario and the northeastern U.S.

Dozens of big-game species from Africa, Asia, and Europe also have been introduced to the United States. Perhaps the first was the wild boar, progenitor of the domestic pig. It was brought to New Hampshire and the Great Smoky Mountains as a sport animal in the early 1900s. Those and subsequent releases generally resulted in interbreeding with more common feral pigs, descended from domestic

Norway rat suckles week-old offspring

stock that had escaped their pens. Today the hunting of alien, big-game mammals is centered in Texas, with hunters paying thousands of dollars each to obtain a single trophy "Texotic." Four introduced deer species, four antelope species, the true wild goat (separate from the domestic goat), the ibex, and the North African Barbary sheep all have established wild-living herds here. Most still have restricted ranges, but the Barbary sheep now occurs in remote and rugged country from Texas to southern California, at times beyond restrictive fences. It may be usurping the habitat of native desert bighorn sheep, which are in serious decline.

Yet another group of aliens contains fully wild species that accidentally found suitable niches in North America. For example, rhesus monkeys from Asia now occur along Florida's Silver Springs River near Ocala, probably brought there as a tourist attraction in the 1930s. The nutria may be the best example of a naturalized

Monkey loose in Florida

wild mammal on our continent. This semiaquatic native of South American marshes, lake edges, and sluggish streams has spread throughout the southeastern U.S. as far west as trans-Pecos Texas. Most nutria populations here are descended from 20 individuals brought to Louisiana in 1938. Part of the captive colony escaped during a hurricane. By the late 1950s there were an estimated 20 million nutria in Louisiana alone; by 1962 they had replaced muskrats as the state's leading fur animal. During the 1976-1977 season, 1,890,853 nutria pelts from Louisiana brought in an average price of $5.25 apiece. Fifteen years later, the take declined to only 240,229 skins at an average price of $3.06 each.

Clearly, trapping nutria and hunting "Texotics" can be profitable to some. But even when the economics are good, such values do not compare to the actual and potential threats posed by introduced species to native wildlife and habitats. Wild horses, wild burros, and certain breeds of feral pigs that date back to the Spanish conquistadores can be of great historical interest. A legitimate case might be made for maintaining modest herds of these animals, under controlled conditions.

North America's entire environmental context today is dominated by yet another category of aliens, a fifth group, one that represents the term "alien" in the best Hollywood tradition. It consists of creatures that have spread throughout their world and even beyond. Their destruction of native wildlife and habitat around the globe far exceeds the lesser ravages of all other species combined, for these particular aliens have learned to alter nature, at times drastically. They alone are ultimately responsible for placing nearly half the world's mammals in jeopardy of extinction. These mammals truly alien to North America are, of course, ourselves. ■

Late arrivals to North America: a burro and a wild horse foal

BY RONALD M. NOWAK

These are the mammals native to the Nearctic Zoogeographical Realm, one of the world's six major biogeographical regions. Each such realm supports its own distinctive and indigenous forms of life. The Nearctic includes all of North America as far south as the highlands of central Mexico (Mexico City). Not included on this list are species whose ranges consist mainly of tropical areas near the Mexican coasts. C, S, N, W, and E are the respective symbols for central, southern, northern, western, and eastern. A bullet (•) denotes an introduced (alien) species.

ORDER DIDELPHIMORPHIA
OPOSSUMS

Family Didelphidae
New World Opossums

Didelphis virginiana, Virginia Opossum: S Ont. to Costa Rica

ORDER INSECTIVORA
INSECTIVORES

Family Soricidae   Shrews
Sorex yukonicus, Alaska Tiny Shrew: C Alas.
S. tundrensis, Tundra Shrew: Siberia, Alas., NW Can.
S. arcticus, Arctic Shrew: Alas., Can., NC U.S.
S. ugyunak, Barren Ground Shrew: Alas. to Hudson Bay
S. pribilofensis, Pribilof Shrew: St. Paul Island (Alas.)
S. haydeni, Hayden's Shrew: SC Can., NC U.S.
S. cinereus, Masked Shrew: Can., N U.S.
S. fontinalis, Maryland Shrew: Penn., West Va., Md., Del.
S. lyelli, Mt. Lyell Shrew: EC Calif.
S. preblei, Preble's Shrew: Oreg. to Mont. and W Colo.
S. milleri, Carmen Mountain Shrew: NE Mex.
S. longirostris, Southeastern Shrew: SE U.S.
S. vagrans, Vagrant Shrew: SW Can., W U.S.
S. oreopolus, Mexican Long-tailed Shrew: C Mex.
S. sonomae, Fog Shrew: SW Oreg., NW Calif.
S. bairdii, Baird's Shrew: NW Oreg.
S. monticolus, Dusky Shrew: Alas. to NW Mex.
S. pacificus, Pacific Shrew: W Oreg.
S. macrodon, Large-toothed Shrew: E Mex.
S. ornatus, Ornate Shrew: Calif., N Baja Calif.
S. tenellus, Inyo Shrew: E Calif., SW Nev.
S. nanus, Dwarf Shrew: Mont. to New Mex.
S. palustris, Water Shrew: SE Alas. to New Mex. and Appalachian Mtns.
S. alaskanus, Glacier Bay Water Shrew: SE Alas.
S. bendirii, Pacific Water Shrew: SW Brit. Col. to NW Calif.
S. fumeus, Smoky Shrew: SE Can., NE U.S.
S. gaspensis, Gaspé Shrew: SE Quebec, N New Bruns.
S. dispar, Long-tailed or Rock Shrew: Nova Scotia to E Tenn.
S. trowbridgii, Trowbridge's Shrew: SW Brit. Col. to C Calif.
S. merriami, Merriam's Shrew: W U.S.
S. arizonae, Arizona Shrew: SE Ariz., SW New Mex., NW Mex.
S. saussurei, Saussure's Shrew: C Mex. to Guatemala
S. emarginatus, Jalisco Long-tailed Shrew: WC Mex.
S. hoyi, Pygmy Shrew: Alas., Can., N U.S.
Cryptotis mexicana, Mexican Small-eared Shrew: E and S Mex.
C. goldmani, Goldman's Small-eared Shrew: C Mex. to Guatemala
C. parva, Least Shrew: SE Ont., E and S U.S., NE Mex. to Panama
Blarina brevicauda, Northern Short-tailed Shrew: S Sask. to Nova Scotia and Ga.
B. carolinensis, Southern Short-tailed Shrew: SE U.S.
B. hylophaga, Elliot's Short-tailed Shrew: S Neb. to E Tex.
Notiosorex crawfordi, Desert Shrew: SW U.S. to C Mex.

Family Talpidae   Moles
Neurotrichus gibbsii, Shrew-mole: S Brit. Col. to C Calif.
Parascalops breweri, Hairy-tailed Mole: SE Can., NE U.S.
Scalopus aquaticus, Eastern Mole: S Ont., C and E U.S.
Scapanus townsendii, Townsend's Mole: SW Brit. Col. to NW Calif.
S. orarius, Coast Mole: SW Brit. Col. to W Idaho and NW Calif.
S. latimanus, Broad-footed Mole: C Oreg. to N Baja Calif.
Condylura cristata, Star-nosed Mole: E Can., E U.S.

ORDER CHIROPTERA   BATS

Family Mormoopidae
Mustached Bats
Mormoops megalophylla, Leaf-chinned or Ghost-faced Bat: SW U.S. to S.A.

Family Phyllostomidae
American Leaf-nosed Bats
Desmodus rotundus, Common Vampire Bat: N Mex. to S.A.
Diphylla ecaudata, Hairy-legged Vampire Bat: N Mex. to S.A.
Macrotus californicus, California Leaf-nosed Bat: SW U.S., NW Mex.
Leptonycteris nivalis, Mexican Long-nosed Bat: SE Ariz. and S Tex. to S.A.
L. curasoae, Southern Long-nosed Bat: S Ariz. and SW New Mex. to S.A.
Choeronycteris mexicana, Hog-nosed or Long-tongued Bat: SW U.S. to Honduras
Dermanura azteca, Highland Fruit Bat: C Mex. to Panama
Artibeus hirsutus, Hairy Fruit Bat: W Mex.

Family Natalidae   Funnel-eared Bats
Natalus stramineus, Greater Funnel-eared Bat: N Mex. to S.A.

Family Vespertilionidae
Vespertilionid Bats
Myotis keenii, Keen's Myotis: SE Alas., Brit. Col., W Wash.
M. septentrionalis, Northern Myotis: C and SE Can., NC and E U.S.
M. evotis, Long-eared Myotis: SW Can., W U.S., Baja Calif.
M. auriculus, Southwestern Myotis: SW U.S. to Guatemala
M. thysanodes, Fringed Myotis: S Brit. Col. to S Mex.
M. ciliolabrum, Western Small-footed Myotis: SW Can. to SC Mex.
M. californicus, California Myotis: SE Alas. to Guatemala
M. leibii, Small-footed Myotis: SE Can., E U.S.
M. sodalis, Indiana or Social Myotis: E U.S.
M. planiceps, Flat-headed Myotis: NE Mex.
M. lucifugus, Little Brown Myotis: Alas., Can., U.S., N Mex.
M. velifer, Cave Myotis: SW U.S. to Honduras
M. grisescens, Gray Myotis: C and SE U.S.
M. yumanensis, Yuma Myotis: Brit. Col. to C Mex.
M. austroriparius, Southeastern Myotis: SE U.S.
M. volans, Long-legged Myotis: SE Alas. to C Mex.
M. vivesi, Mexican Fishing Bat or Fish-eating Bat: W Mex.
Lasionycteris noctivagans, Silver-haired Bat: S Can., U.S., N Mex.
Pipistrellus subflavus, Eastern Pipistrelle: Minn. and Nova Scotia to Honduras and Fla.
P. hesperus, Western Pipistrelle: S Wash. to C Mex.
Nycticeius humeralis, Evening Bat: E U.S., NE Mex.
Eptesicus fuscus, Big Brown Bat: S Can. to S.A.
Rhogeessa alleni, Allen's Yellow Bat: C Mex.
Lasiurus intermedius, Northern Yellow Bat: E U.S., Mex. to Honduras
L. ega, Southern Yellow Bat: S Tex. to S.A.
L. xanthinus, Western Yellow Bat: SW U.S. to C Mex.
L. cinereus, Hoary Bat: Can. to S.A., Hawaii
L.seminolus, Seminole Bat: SE U.S.
L. borealis, Eastern Red Bat: SE Can. to N Mex.
L. blossevillii, Western Red Bat: SW Can. to S.A.
Corynorhinus mexicanus, Mexican Big-eared Bat: Mex.
C. rafinesquii, Rafinesque's Big-eared Bat: SE U.S.
C. townsendii, Townsend's Big-eared Bat: S Brit. Col. to S Mex.; parts of E U.S.
Idionycteris phyllotis, Allen's Big-eared Bat: S Utah to C Mex.
Euderma maculatum, Spotted Bat: S Brit. Col. to N Mex.
Antrozous pallidus, Pallid Bat: S Brit. Col. to C Mex., Cuba

Family Molossidae
Free-tailed Bats
Tadarida brasiliensis, Mexican or Brazilian Free-tailed Bat: U.S. to S.A.
Nyctinomops femorosaccus, Pocketed Free-tailed Bat: SW U.S. to S.A.
N. macrotis, Big Free-tailed Bat: SW U.S. to S.A.
Eumops underwoodi, Underwood's Mastiff Bat: S Ariz. to Nicaragua
E. glaucinus, Wagner's Mastiff Bat: S Fla., C Mex. to S.A., West Indies
E. perotis, Western Mastiff Bat: SW U.S. to S.A., Cuba
Molossus molossus, Velvety Free-tailed Bat: Fla. Keys

ORDER PRIMATES   PRIMATES

Family Cercopithecidae
Old World Monkeys
•*Macaca mulatta,* Rhesus Macaque (S and E Asia) C Fla.
•*M. fuscata,* Japanese Macaque: (Japan) S Fla.

Family Hominidae
Humans
Homo sapiens, People: Worldwide

ORDER XENARTHRA
XENARTHRANS

Family Dasypodidae   Armadillos
Dasypus novemcinctus, Common Long-nosed or Nine-banded Armadillo: S U.S. to S.A.

ORDER LAGOMORPHA
LAGOMORPHS

Family Ochotonidae   Pikas
Ochotona collaris, Collared Pika: E Alas., NW Can.
O. princeps, American Pika: SW Can., W U.S.

Family Leporidae   Hares and Rabbits
Romerolagus diazi, Volcano Rabbit: C Mex.
Brachiolagus idahoensis, Pygmy Rabbit: NW U.S.
Sylvilagus bachmani, Brush Rabbit: W Oreg. to Baja Calif.
S. aquaticus, Swamp Rabbit: SC U.S.
S. palustris, Marsh Rabbit: SE U.S.
S. transitionalis, New England Cottontail: New Eng., E N.Y.
S. obscurus, Appalachian Cottontail: W N.Y. to N Ala.
S. floridanus, Eastern Cottontail: S Can. to S.A.
S. nuttallii, Nuttall's or Mountain Cottontail: SW Can., W U.S.
S. audubonii, Desert Cottontail: W U.S. to C Mex.
•*Oryctolagus cuniculus,* European Rabbit (SW Europe): San Juan Islands, Wash.
•*Lepus europaeus,* European Hare (Eurasia): SE Can., NE U.S.
Lepus americanus, Snowshoe Hare: Alas., Can., N U.S.
L. othus, Alaskan Hare: Alas., E Siberia
L. arcticus, Arctic Hare: N. Can., Greenland

L. townsendii, White-tailed Jackrabbit: SW Can., W U.S.

L. californicus, Black-tailed Jackrabbit: W and C U.S., N Mex.

L. callotis, White-sided Jackrabbit: SW New Mex. to C Mex.

L. alleni, Antelope Jackrabbit: S Ariz., NW Mex.

ORDER RODENTIA
RODENTS

Family Aplodontidae
Mountain Beaver

Aplodontia rufa, Mountain Beaver: SW Brit. Col. to C Calif.

Family Sciuridae   Squirrels

Tamias striatus, Eastern Chipmunk: SE Can., E U.S.

T. merriami, Merriam's Chipmunk: C and S Calif., N Baja Calif.

T. obscurus, California Chipmunk: S Calif., Baja Calif.

T. alpinus, Alpine Chipmunk: C Calif.

T. palmeri, Palmer's Chipmunk: S Nev.

T. bulleri, Buller's Chipmunk: WC Mex.

T. sonomae, Sonoma Chipmunk: NW Calif.

T. townsendii, Townsend's Chipmunk: SW Brit Col. to W Oreg.

T. siskiyou, Siskiyou Chipmunk: SW Oreg., NW Calif.

T. senex, Allen's Chipmunk: C Oreg., N Calif.

T. ochrogenys, Yellow-cheeked Chipmunk: NW Calif.

T. quadrimaculatus, Long-eared Chipmunk: EC Calif., N Nev.

T. canipes, Gray-footed Chipmunk: SC New Mex., SW Tex.

T. durangae, Durango Chipmunk: NW Mex.

T. cinereicollis, Gray-collared Chipmunk: EC Ariz., SW New Mex.

T. amoenus, Yellow-pine Chipmunk: C Brit. Col. to EC Calif. and W Wyo.

T. umbrinus, Uinta Chipmunk: Wyo. to SE Calif.

T. ruficaudus, Red-tailed Chipmunk: SW Can., NW U.S.

T. speciosus, Lodgepole Chipmunk: Calif., W Nev.

T. minimus, Least Chipmunk: Can., W U.S.

T. panamintinus, Panamint Chipmunk: SE Calif., SW Nev.

T. quadrivittatus, Colorado Chipmunk: Colo. and adjacent areas

T. rufus, Hopi Chipmunk: E Utah, W Colo., NC Ariz.

T. dorsalis, Cliff Chipmunk: S Idaho to N Mex.

Marmota broweri, Alaska Marmot: Alas.

M. monax, Woodchuck: Alas., Can., N and E U.S.

M. flaviventris, Yellow-bellied Marmot: SW Can., W U.S.

M. caligata, Hoary Marmot: Alas. to Wash. and W Mont.

M. olympus, Olympic Marmot: NW Wash.

M. vancouverensis, Vancouver Marmot: Vancouver Island (Brit. Col.)

Ammospermophilus harrisii, Harris' Antelope Squirrel: SW U.S., NW Mex.

A. leucurus, White-tailed Antelope Squirrel: SE Oreg. to Baja Calif.

A. interpres, Texas Antelope Squirrel: SC New Mex., W Tex., NC Mex.

A. nelsoni, Nelson's Antelope Squirrel: C Calif.

Spermophilus parryii, Arctic Ground Squirrel: Alas., NW Can., NE Siberia

S. columbianus, Columbian Ground Squirrel: SW Can., NW U.S.

S. townsendii, Townsend's Ground Squirrel: SC Wash.

S. canus, Columbia Plateau Ground Squirrel: E Oreg. NW Nev., WC Idaho

S. mollis, Great Basin Ground Squirrel: C Wash. to EC Calif. and W Utah

S. washingtoni, Washington Ground Squirrel: SE Wash., NE Oreg.

S. armatus, Uinta Ground Squirrel: SW Mont., SE Idaho, W Wyo., N Utah

S. beldingi, Belding's Ground Squirrel: E Oreg. to NW Utah and NE Calif.

S. richardsonii, Richardson's Ground Squirrel: C Alberta and Mont. to W Minn.

S. elegans, Wyoming Ground Squirrel: SE Mont. and N Nev. to N Colo.

S. brunneus, Idaho Ground Squirrel: SW Idaho

S. tridecemlineatus, Thirteen-lined Ground Squirrel: C Alberta to Ohio and S Tex.

S. mexicanus, Mexican Ground Squirrel: N Tex. to C Mex.

S. spilosoma, Spotted Ground Squirrel: North Dak. to N Mex.

S. perotensis, Perote Ground Squirrel: C Mex.

S. franklinii, Franklin's Ground Squirrel: Alberta to Ohio

S. variegatus, Rock Squirrel: SW U.S. to C Mex.

S. beecheyi, California Ground Squirrel: S Wash. to N Baja Calif.

S. atricapillus, Baja California Rock Squirrel: C Baja Calif.

S. mohavensis, Mohave Ground Squirrel: SC Calif.

S. tereticaudus, Round-tailed Ground Squirrel: SW U.S., NW Mex.

S. lateralis, Golden-mantled Ground Squirrel: SW Can., W U.S.

S. saturatus, Cascade Golden-mantled Ground Squirrel: SE Brit. Col., C Wash.

S. madrensis, Sierra Madre Mantled Ground Squirrel: NW Mex.

Cynomys ludovicianus, Black-tailed Prairie Dog: S Sask. to N Mex.

C. mexicanus, Mexican Prairie Dog: NE Mex.

C. leucurus, White-tailed Prairie Dog: SC Mont., W Wyo., NE Utah, NW Colo.

C. parvidens, Utah Prairie Dog: SC Utah

C. gunnisoni, Gunnison's Prairie Dog: SW U.S.

Sciurus carolinensis, Gray Squirrel: SE Can., E U.S.

S. aureogaster, Mexican Gray Squirrel: Mex., Guatemala

S. colliaei, Collie's Squirrel: W Mex.

S. niger, Fox Squirrel: C and E U.S., NE Mex.

S. oculatus, Peters' Squirrel: C Mex.

S. alleni, Allen's Squirrel: NE Mex.

S. nayaritensis, Nayarit Squirrel: SE Ariz., W Mex.

S. arizonensis, Arizona Gray Squirrel: Ariz., W New Mex., NW Mex.

S. griseus, Western Gray Squirrel: NC Wash. to S Calif.

S. aberti, Abert's Squirrel: SW U.S., NW Mex.

Tamiasciurus hudsonicus, Red Squirrel: Alas., Can., Rocky Mts., NE U.S.

T. douglasii, Douglas' Squirrel: SW Brit. Col. to C Calif.

T. mearnsi, Mearns' Squirrel: N Baja Calif.

Glaucomys volans, Southern Flying Squirrel: SE Can., E U.S., Mex. to Honduras

G. sabrinus, Northern Flying Squirrel: Can., N and W U.S., Appalachian Mts.

Family Geomyidae
Pocket Gophers

Thomomys talpoides, Northern Pocket Gopher: SW Can., W U.S.

T. idahoensis, Idaho Pocket Gopher: SW Mont., SE Idaho, W Wyo., NE Utah

T. clusius, Wyoming Pocket Gopher: SC Wyo.

T. mazama, Western Pocket Gopher: W Wash. to NC Calif.

T. monticola, Mountain Pocket Gopher: NE Calif., W Nev.

T. bulbivorus, Camas Pocket Gopher: NW Oreg.

T. bottae, Botta's Pocket Gopher: SW Oreg. and C Colo. to C Mex.

T. townsendii, Townsend's Pocket Gopher: SE Oreg., S Idaho, NE Calif., N Nev.

T. umbrinus, Southern Pocket Gopher: SW U.S., N Mex.

Geomys bursarius, Northern Plains Pocket Gopher: S Manitoba to E Kan.

G. lutescens, Southern Plains Pocket Gopher: E Wyo. to NC Tex.

G. texensis, Edwards Plateau Pocket Gopher: C Tex.

G. knoxjonesi, Jones' Pocket Gopher: SE New Mex., W Tex.

G. breviceps, Baird's Pocket Gopher: SC U.S.

G. arenarius, Desert Pocket Gopher: SC New Mex., W Tex., NE Mex.

G. attwateri, Attwater's Pocket Gopher: SC Tex.

G. personatus, Texas Pocket Gopher: S Tex., NE Mex.

G. tropicalis, Tropical Pocket Gopher: NE Mex.

G. pinetis, Southeastern Pocket Gopher: S Ala., S Ga., N Fla.

Zygogeomys trichopus, Michoacán Pocket Gopher: C Mex.

Pappogeomys bulleri, Buller's Pocket Gopher: WC Mex.

P. alcorni, Alcorn's Pocket Gopher: WC Mex.

Cratogeomys castanops, Yellow-faced Pocket Gopher: SW Kan. to C Mex.

C. merriami, Merriam's Pocket Gopher: EC Mex.

C. neglectus, Querétaro Pocket Gopher: C Mex.

C. fumosus, Smoky Pocket Gopher: C Mex.

C. tylorhinus, Taylor's Pocket Gopher: C Mex.

C. zinseri, Zinser's Pocket Gopher: WC Mex.

C. gymnurus, Llano Pocket Gopher: WC Mex.

Family Heteromyidae   Heteromyids

Liomys irroratus, Mexican Spiny Pocket Mouse: S Tex., Mex.

L. pictus, Painted Spiny Pocket Mouse: W and S Mex., Guatemala

Perognathus fasciatus, Olive-backed Pocket Mouse: S Sask. to S Colo.

P. flavescens, Plains Pocket Mouse: E Utah to Minn. and NC Mex.

P. parvus, Great Basin Pocket Mouse: S Brit. Col. to SC Calif. and Utah

P. alticolus, White-eared Pocket Mouse: SC Calif.

P. flavus, Silky Pocket Mouse: SW U.S. to C Mex.

P. merriami, Merriam's Pocket Mouse: E New Mex. to NC Tex. and NW Mex.

P. longimembris, Little Pocket Mouse: SE Oreg. to NW Mex.

P. amplus, Arizona Pocket Mouse: Ariz., NW Mex.

P. inornatus, San Joaquin Pocket Mouse: C Calif.

Chaetodipus baileyi, Bailey's Pocket Mouse: SW U.S., NW Mex.

C. penicillatus, Desert Pocket Mouse: SW U.S., NW Mex.

C. eremicus, Chihuahuan Desert Pocket Mouse: SW U.S., NC Mex.

C. formosus, Long-tailed Pocket Mouse: SW U.S., Baja Calif.

C. artus, Narrow-skulled Pocket Mouse: NW Mex.

C. lineatus, Lined Pocket Mouse: C Mex.

C. nelsoni, Nelson's Pocket Mouse: SE New Mex., W Tex., NC Mex.

C. pernix, Sinaloan Pocket Mouse: NW Mex.

C. goldmani, Goldman's Pocket Mouse: NW Mex.

C. intermedius, Rock Pocket Mouse: SW U.S., NW Mex.

C. arenarius, Little Desert Pocket Mouse: Baja Calif.

C. fallax, San Diego Pocket Mouse: S Calif., Baja Calif., Cerros Island (NW Baja Calif.)

C. spinatus, Spiny Pocket Mouse: S Calif., Baja Calif.

C. californicus, California Pocket Mouse: S Calif., Baja Calif.

C. hispidus, Hispid Pocket Mouse: N Dak. to C Mex.

Microdipodops megacephalus, Dark Kangaroo Mouse: Great Basin region (W U.S.)

M. pallidus, Pale Kangaroo Mouse: W Nev., E Calif.

Dipodomys ordii, Ord's Kangaroo Rat: SC Can. to C Mex.

D. compactus, Gulf Coast Kangaroo Rat: S Tex., NE Mex.

D. microps, Chisel-toothed Kangaroo Rat: Oreg. to N Ariz.

D. panamintinus, Panamint Kangaroo Rat: W Nev., S Calif.

D. stephensi, Stephens' Kangaroo Rat: SW Calif.

D. elephantinus, Big-eared Kangaroo Rat: WC Calif.

D. venustus, Narrow-faced Kangaroo Rat: W Calif.

D. agilis, Agile Kangaroo Rat: SW Calif.

D. simulans, Dulzura Kangaroo Rat: SW Calif., Baja Calif.

D. heermanni, Heermann's Kangaroo Rat: C Calif.

D. californicus, California Kangaroo Rat: SC Oreg., N Calif.

D. gravipes, San Quintín Kangaroo Rat: NW Baja Calif.

D. ingens, Giant Kangaroo Rat: SW Calif.

D. spectabilis, Banner-tailed Kangaroo Rat: SW U.S., N Mex.
D. nelsoni, Nelson's Kangaroo Rat: N Mex.
D. elator, Texas Kangaroo Rat: SW Okla., NC Tex.
D. phillipsii, Phillips' Kangaroo Rat: C Mex.
D. merriami, Merriam's Kangaroo Rat: SW U.S., N Mex.
D. insularis, San José Kangaroo Rat: San José Island (SE Baja Calif.)
D. margaritae, Margarita Kangaroo Rat: Margarita Island (SW Baja Calif.)
D. nitratoides, Fresno Kangaroo Rat: C Calif.
D. deserti, Desert Kangaroo Rat: NW Nev. to NW Mex.

Family Castoridae   Beavers

Castor canadensis, Beaver: Alas., Can., U.S., N Mex.

Family Muridae   Rats and Mice

Nelsonia neotomodon, Diminutive Woodrat: WC Mex.
N. goldmani, Goldman's Diminutive Woodrat: C Mex.
Neotoma floridana, Eastern Woodrat: S. Dak. and C Tex. to Fla.
N. magister, Appalachian Woodrat: S N.Y. to N Ala.
N. micropus, Southern Plains Woodrat: S Kan. to NE Mex.
N. albigula, White-throated Woodrat: SW U.S. to C Mex.
N. nelsoni, Nelson's Woodrat: EC Mex.
N. palatina, Balaños Woodrat: WC Mex.
N. varia, Turner Island Woodrat: Turner Island (NW Mex.)
N. lepida, Desert Woodrat: C Calif. and S Idaho to Baja Calif.
N. devia, Painted Desert Woodrat: W Ariz.
N. bryanti, Bryant's Woodrat: Cerros Island (NW Baja Calif.)
N. anthonyi, Anthony's Woodrat: Todos Santos Island (NW Baja Calif.)
N. martinensis, San Martín Island Woodrat: San Martín Island (NW Baja Calif.)
N. bunkeri, Coronados Island Woodrat: Coronados Island (SE Baja Calif.)
N. stephensi, Stephen's Woodrat: SW U.S.
N. goldmani, Goldman's Woodrat: NC Mex.
N. mexicana, Mexican Woodrat: SW U.S. to Honduras
N. angustapalata, Tamaulipan Woodrat: NE Mex.
N. fuscipes, Dusky-footed Woodrat: W Oreg. to N Baja Calif.
N. cinerea, Bushy-tailed Woodrat: NW Can. to N. Dak. and N Ariz.
Ochrotomys nuttalli, Golden Mouse: SC and SE U.S.
Baiomys taylori, Northern Pygmy Mouse: SW U.S. to C Mex.
Onychomys torridus, Southern Grasshopper Mouse: SW U.S. to N Mex.
O. leucogaster, Northern Grasshopper Mouse: E Wash. to N Mex.
O. arenicola, Eastern Grasshopper Mouse: SW New Mex., W Tex., NC Mex.
Neotomodon alstoni, Volcano Mouse: C Mex.
Podomys floridanus, Florida Mouse: Fla.
Peromyscus californicus, California Mouse: C Calif. to N Baja Calif.
P. eremicus, Cactus Mouse: SW U.S., N Mex.
P. guardia, Angel Island Mouse: Angel Island (NE Baja Calif.)
P. interparietalis, San Lorenzo Mouse: Salsipuedes and San Lorenzo Islands (NE Baja Calif.)
P. dickeyi, Dickey's Mouse: Tortuga Island (EC Baja Calif.)
P. pseudocrinitus, Coronados Island Mouse: Coronados Island (SE Baja Calif.)
P. eva, Eva's Desert Mouse: S Baja Calif.
P. caniceps, Monserrate Mouse: Monserrate Island (SE Baja Calif.)
P. merriami, Merriam's Mouse: S Ariz., NW Mex.
P. pembertoni, Pemberton's Mouse: San Pedro Nolasco Island (NW Mex.)
P. hooperi, Hooper's Mouse: NE Mex.
P. crinitus, Canyon Mouse: W U.S., NW Mex.
P. maniculatus, Deer Mouse: SE Alas., Can., U.S., Mex.
P. polionotus, Oldfield Mouse: SE U.S.
P. sejugis, Santa Cruz Island Mouse: Santa Cruz

and San Diego Islands (SE Baja Calif.)
P. oreas, Mount Baker Mouse: SW Brit. Col.,W Wash.
P. sitkensis, Sitka Mouse: SE Alas., Queen Charlotte Islands (Brit. Col.)
P. melanotis, Black-eared Mouse: SE Ariz., N and C Mex.
P. slevini, Slevin's Mouse: Santa Catalina Island (SE Baja Calif.)
P. nesodytes, San Miguel Mouse: San Miguel Island (S Calif., extinct)
P. leucopus, White-footed Mouse: S Can. to S Mex.
P. gossypinus, Cotton Mouse: SE U.S.
P. spicilegus, Jalisco Mouse: W C Mex.
P. boylii, Brush Mouse: SW U.S. to C Mex.
P. stephani, San Esteban Island Mouse: San Esteban Island (NE Baja Calif.)
P. attwateri, Texas Mouse: SC U.S.
P. pectoralis, White-ankled Mouse: S Okla. to C Mex.
P. polius, Chihuahuan Mouse: NC Mex.
P. truei, Piñon Mouse: SW Oreg. to N Baja Calif. and W Tex.
P. gratus, Osgood's Mouse: C New Mex. to S Mex.
P. difficilis, Southern Rock Mouse: Mex.
P. nasutus, Northern Rock Mouse: Colo. to NC Mex.
P. melanophrys, Plateau Mouse: Mex.
P. furvus, Blackish Deer Mouse: EC Mex.
P. ochraventer, El Carrizo Deer Mouse: C Mex.
Reithrodontomys montanus, Plains Harvest Mouse: S. Dak. to NW Mex.
R. burti, Sonoran Harvest Mouse: NW Mex.
R. humulis, Eastern Harvest Mouse: SE U.S.
R. megalotis, Western Harvest Mouse: SW Can. to Ind. and S Mex.
R. zacatecae, Zacatecas Harvest Mouse: W Mex.
R. raviventris, Salt Marsh Harvest Mouse: San Francisco Bay area
R. chrysopsis, Volcano Harvest Mouse: C Mex.
R. sumichrasti, Sumichrast's Harvest Mouse: C Mex. to Panama
R. fulvescens, Fulvous Harvest Mouse: SC U.S. to Nicaragua
R. hirsutus, Hairy Harvest Mouse: WC Mex.
Oryzomys palustris, Marsh Rice Rat: SE U.S.
O. couesi, Coues' Rice Rat: S Tex. to S.A.
O. argentatus, Silver Rice Rat: Fla. Keys
Sigmodon hispidus, Hispid Cotton Rat: S U.S. to S.A.
S. alleni, Allen's Cotton Rat: W Mex.
S. leucotis, White-eared Cotton Rat: C Mex.
S. ochrognathus, Yellow-nosed Cotton Rat: SW U.S., N Mex.
S. arizonae, Arizona Cotton Rat: SE Calif., Ariz., NW Mex.
S. fulviventer, Tawny-bellied Cotton Rat: SE U.S. to C Mex.
Clethrionomys rutilus, Northern Red-backed Vole: Alas., NW Can., N Eurasia
C. gapperi, Southern Red-backed Vole: S Can., N U.S., Rocky Mtns.
C. californicus, California Red-backed Vole: W Oreg., N Calif.
Arborimus albipes, White-footed Vole: W Oreg., NW Calif.
A. longicaudus, Red Tree Vole: W Oreg.
A. pomo, Sonoma Tree Vole: NW Calif.
Phenacomys intermedius, Western Heather Vole: SW Can., W U.S.
P. ungava, Eastern Heather Vole: Can., N Minn.
Microtus pinetorum, Pine or Woodland Vole: S Ont., E U.S.
M. quasiater, Jalapan Pine Vole: C Mex.
M. ochrogaster, Prairie Vole: Alberta to West Va. and New Mex.
M. richardsoni, Water Vole: SW Can., NW U.S.
M. xanthognathus, Yellow-cheeked Vole: Alas., W Can.
M. chrotorrhinus, Rock Vole: SE Can., NE U.S.
M. longicaudus, Long-tailed Vole: E Alas. to New Mex.
M. coronarius, Coronation Island Vole: SE Alas.
M. oeconomus, Tundra Vole: Alas., NW Can., Eurasia
M. miurus, Singing Vole: Alas., NW Can.
M. abbreviatus, Insular Vole: Hall and St. Matthew Islands (Alas.)
M. pennsylvanicus, Meadow Vole: Alas., Can.,

U.S., NC Mex.
M. nesophilus, Gull Island Vole: Gull Islands (N.Y., extinct)
M. breweri, Beach Vole: Muskeget Island (Mass.)
M. townsendii, Townsend's Vole: SW Brit. Col. to NW Calif.
M. montanus, Montane Vole: SW Can., W U.S.
M. canicaudus, Gray-tailed Vole: SW Wash., W Oreg.
M. oregoni, Creeping Vole: SW Brit Col. to NW Calif.
M. californicus, California Vole: SW Oreg. to N Baja Calif.
M. mexicanus, Mexican Vole: Mex.
M. mogollonensis, Mogollon Vole: SW U.S.
Lemmiscus curtatus, Sagebrush Vole: SW Can., W U.S.
Neofiber alleni, Round-tailed Muskrat: S Ga., Fla.
Ondatra zibethicus, Muskrat: Alas., Can., U.S., N Baja Calif.
Dicrostonyx groenlandicus, Collared Lemming: Alas., NW and NC Can., Greenland
D. hudsonius, Labrador Collared Lemming: NE Can.
Lemmus trimucronatus, Brown Lemming: Alas., W and NC Can.
Synaptomys cooperi, Southern Bog Lemming: SE Can., C and NE U.S.
S. borealis, Northern Bog Lemming: Alas., Can.
•*Rattus rattus*, Black Rat: Worldwide in association with people
•*R. norvegicus*, Norway Rat: Worldwide in association with people
•*Mus musculus*, House Mouse: Worldwide in association with people

Family Zapodidae   Jumping Mice

Zapus hudsonius, Meadow Jumping Mouse: S Alas., Can., U.S.
Z. princeps, Western Jumping Mouse: SW Can., W U.S.
Z. trinotatus, Pacific Jumping Mouse: SW Brit. Col. to NW Calif.
Napaeozapus insignis, Woodland Jumping Mouse: SE Can., NE U.S.

Family Erethizontidae
New World Porcupines

Erethizon dorsatum, Porcupine: Alas., Can., U.S., N Mex.

Family Myocastoridae   Nutria

•*Myocastor coypus*, Nutria: (S.A.) SE U.S.

ORDER CETACEA
CETACEANS

Family Ziphiidae   Beaked Whales

Berardius bairdii, Baird's Beaked Whale: Pacific coast
Ziphius cavirostris, Goose-beaked Whale: Atlantic and Pacific coasts
Hyperoodon ampullatus, Northern Bottlenose Whale: Atlantic coast
Mesoplodon peruvianus, Eastern Pacific Beaked Whale: Pacific coast
M. hectori, Hector's Beaked Whale: Pacific coast
M. mirus, True's Beaked Whale: Atlantic coast
M. europaeus, Gervais' Beaked Whale: Atlantic coast
M. ginkgodens, Japanese Beaked Whale: Pacific coast
M. carlhubbsi, Moore's Beaked Whale: Pacific coast
M. stejnegeri, North Pacific Beaked Whale: Pacific coast
M. bidens, North Atlantic Beaked Whale: Atlantic coast
M. densirostris, Tropical Beaked Whale: Atlantic and Pacific coasts

Family Physeteridae
Sperm Whales

Kogia breviceps, Pygmy Sperm Whale: Atlantic and Pacific coasts
K. simus, Dwarf Sperm Whale: Atlantic and Pacific coasts
Physeter catodon, Sperm Whale: Atlantic and

Pacific coasts

Family Monodontidae
Monodontids

Delphinapterus leucas, White Whale or Beluga: Arctic, Atlantic, and Pacific coasts
Monodon monoceros, Narwhal: Arctic, Atlantic, and Pacific coasts

Family Delphinidae   Delphinids

Steno bredanensis, Rough-toothed Dolphin: Atlantic and Pacific coasts
Lagenorhynchus albirostris, White-beaked Dolphin: Atlantic coast
L. acutus, Atlantic White-sided Dolphin: Atlantic coast
L. obliquidens, Pacific White-sided Dolphin: Pacific coast
Grampus griseus, Grampus or Risso's Dolphin: Atlantic and Pacific coasts
Tursiops truncatus, Bottlenose Dolphin: Atlantic and Pacific coasts
Stenella longirostris, Spinner Dolphin: Atlantic and Pacific coasts
S. clymene Atlantic Spinner Dolphin: Atlantic coast
S. coeruleoalba, Striped Dolphin: Atlantic and Pacific coasts
S. attenuata, Bridled or Pantropical Spotted Dolphin: Atlantic and Pacific coasts
S. frontalis, Atlantic Spotted Dolphin: Atlantic coast
Delphinus delphis, Common or Saddleback Dolphin: Atlantic and Pacific coasts
D. capensis, Long-beaked Saddleback Dolphin: Pacific coast
Lagenodelphis hosei, Fraser's Dolphin: Atlantic and Pacific coasts
Lissodelphis borealis, Northern Right Whale Dolphin: Pacific coast
Peponocephala electra, Melon-headed Whale: Atlantic coast
Feresa attenuata, Pygmy Killer Whale: Gulf coast
Pseudorca crassidens, False Killer Whale: Atlantic and Pacific coasts
Orcinus orca, Killer Whale: Atlantic and Pacific coasts
Globicephala melas, Long-finned Pilot Whale: Atlantic coast
G. macrorhynchus, Short-finned Pilot Whale: Atlantic and Pacific coasts
Phocoena phocoena, Harbor Porpoise: Atlantic and Pacific coasts
P. sinus, Cochito or Gulf of California Harbor Porpoise: Gulf of California
P. dalli, Dall's Porpoise: Pacific coast

Family Eschrichtiidae
Gray Whale

Eschrichtius robustus, Gray Whale: Pacific coast

Family Balaenopteridae
Rorqual Whales

Balaenoptera acutorostrata, Minke Whale: Atlantic and Pacific coasts
B. brydei, Bryde's Whale: Atlantic and Pacific coasts
B. borealis, Sei Whale: Atlantic and Pacific coasts
B. physalus, Fin Whale: Atlantic and Pacific coasts
B musculus, Blue Whale: Atlantic and Pacific coasts
Megaptera novaeangliae, Humpback Whale: Atlantic and Pacific coasts

Family Balaenidae
Right and Bowhead Whales

Balaena glacialis, Right Whale: Atlantic and Pacific coasts
B. mysticetus, Bowhead Whale: Atlantic and Pacific coasts

ORDER CARNIVORA
CARNIVORES

Family Canidae   Canids

Vulpes vulpes, Red Fox: Alas., Can., U.S., Eurasia

V. velox, Swift Fox: S Alberta to NW Tex.
V. macrotis, Kit Fox: S Oreg. to Baja Calif. and NC Mex.
Alopex lagopus, Arctic Fox: Alas., N Can., Greenland, Iceland, N Eurasia
Urocyon cinereoargenteus, Gray Fox: Oreg. and SE Can. to S.A.
U. littoralis, Insular Gray Fox: Channel Islands (Calif.)
Canis latrans, Coyote: Alas. to Nova Scotia and Panama
C. rufus, Red Wolf: C Tex. to S Penn. and Fla.
C. lupus, Gray Wolf: Alas., Can., Greenland, U.S., Mex., Eurasia
•*C. familiaris,* Domestic Dog: Worldwide in association with people

Family Ursidae   Bears

Ursus americanus, Black Bear: Alas., Can., U.S., N Mex.
U. arctos, Grizzly or Brown Bear: Alas. to Hudson Bay and N Mex., Eurasia
U. maritimus, Polar Bear: Alas., N Can., N Eurasia

Family Procyonidae   Procyonids

Bassariscus astutus, Ringtail: SW Oreg. and E Kan. to S Mex.
Procyon lotor, Raccoon: S Can. to Panama
Nasua narica, Coati: Ariz. to S.A.

Family Mustelidae   Mustelids

Mustela erminea, Ermine: Alas., Can., N U.S., Greenland, Eurasia
M. nivalis, Least Weasel: Alas., Can., N U.S., Eurasia, N Africa
M. frenata, Long-tailed Weasel: S Can. to S.A.
M. vison, Mink: Alas., Can., U.S.
M. macrodon, Sea Mink: New Bruns. to Mass. (extinct)
•*M. putorius,* European Ferret: (Europe) SW U.S.
M. nigripes, Black-footed Ferret: Alberta to Tex.
Martes americana, Marten: Alas., Can., N U.S.
M. pennanti, Fisher: S Can., U.S.
Gulo gulo, Wolverine: Alas., Can., N U.S., N Eurasia
Taxidea taxus, Badger: SW Can. to C Mex.
Spilogale gracilis, Western Spotted Skunk: S Brit. Col. and C Wyo. to Costa Rica
S. putorius, Eastern Spotted Skunk: E Wyo. to Penn. and Fla.
Mephitis mephitis, Striped Skunk: S Can. to N Mex.
M. macroura, Hooded Skunk: Ariz. to Costa Rica
Conepatus mesoleucus, Hog-nosed Skunk: S Colo. and E Tex. to Nicaragua
C. leuconotus, Eastern Hog-nosed Skunk: S Tex., E Mex.
Lontra canadensis, River Otter: Alas., Can., U.S.
Enhydra lutris, Sea Otter: Pacific Coast

Family Felidae   Cats

•*Felis catus,* Domestic Cat: Worldwide in association with people
Lynx canadensis, Lynx: Alas., Can., N U.S.
L. rufus, Bobcat: S Can., U.S., Mex.
Leopardus pardalis, Ocelot: Ariz. and Tex. to S.A.
L. wiedii, Margay, N Mex. to S.A.
Herpailurus yagouaroundi, Jaguarundi: S Ariz. and S Tex. to S.A.
Puma concolor, Mountain Lion or Cougar: Can. to S.A.
Panthera onca, Jaguar: SW U.S. to S.A.

ORDER PINNIPEDIA
PINNIPEDS

Family Otariidae   Eared Seals

Callorhinus ursinus, Northern Fur Seal: Pacific coast
Arctocephalus townsendi, Guadalupe Fur Seal: Pacific coast
Eumetopias jubatus, Northern Sea Lion: Pacific coast
Zalophus californianus, California Sea Lion: Pacific coast

Family Odobenidae   Walrus

Odobenus rosmarus, Walrus: Atlantic and Pacific

coasts, Hudson Bay

Family Phocidae   True Seals

Monachus tropicalis, West Indian Monk Seal: Gulf coast (extinct)
Mirounga angustirostris, Northern Elephant Seal: Pacific coast
Erignathus barbatus, Bearded Seal: Bering Sea, Hudson Bay, Ungava Bay
Cystophora cristata, Hooded Seal: N Atlantic coast
Halichoerus grypus, Gray Seal: Atlantic coast
Phoca groenlandica, Harp Seal: N Atlantic coast, Hudson Bay
P. hispida, Ringed Seal: Arctic, Atlantic, and Pacific coasts
P. largha, Spotted Seal: Arctic coast
P. vitulina, Harbor Seal: Atlantic and Pacific coasts
P. fasciata, Ribbon Seal: Alaskan waters

ORDER SIRENIA
SEA COWS

Family Trichechidae   Manatees

Trichechus manatus, Manatee: Gulf coast, Fla., Caribbean

ORDER PERISSODACTYLA
ODD-TOED UNGULATES

Family Equidae   Equids

•*Equus asinus,* Ass or Burro: (NE Africa) Worldwide
•*E. caballus,* Horse: (Eurasia) Worldwide

ORDER ARTIODACTYLA
EVEN-TOED UNGULATES

Family Suidae   Pigs

•*Sus scrofa,* Pig or Wild Boar: (Eurasia) S U.S.

Family Tayassuidae   Peccaries

Dicotyles tajacu, Collared Peccary: SW U.S. to S.A.

Family Cervidae   Deer

•*Dama dama,* Fallow Deer (Europe, SW Asia) Can., U.S.
•*Axis axis,* Axis Deer: (SC Asia) SW U.S.
•*Cervus unicolor,* Sambar Deer: (SE Asia) S U.S.
•*C. nippon,* Sika Deer: (E Asia) S U.S.
C. elaphus, Elk or Wapiti: S Can., U.S., Eurasia
Odocoileus hemionus, Mule Deer: SE Alas. and SW Can. to N Mex.
O. virginianus, White-tailed Deer: S Can. to S.A.
Alces alces, Moose: Alas., Can., N U.S., Eurasia
Rangifer tarandus, Caribou: Alas., Can., N U.S., Greenland, N Eurasia

Family Antilocapridae
Pronghorn

Antilocapra americana, Pronghorn: SW Can. to N Mex.

Family Bovidae   Bovids

•*Bos taurus,* Domestic Cattle: Worldwide
Bison bison, Bison: W Can., U.S.
•*Boselaphus tragocamelus,* Nilgai: (SC Asia) Tex.
•*Oryx gazella,* Gemsbok: (Africa) SW U.S.
•*Antelope cervicapra,* Blackbuck: (SC Asia) Tex.
Oreamnos americanus, Mountain Goat: SE Alas. to Oreg.
Ovibos moschatus, Muskox: Alas., N Can., Greenland
•*Hemitragus jemlahicus,* Himalayan Tahr: (Asia) SW U.S.
•*Capra hircus,* Domestic Goat: Worldwide
•*C. ibex,* Ibex: (Eurasia) SW U.S.
•*Ammotragus lervia,* Barbary Sheep or Aoudad: (N Africa) SW U.S.
•*Ovis orientalis,* Mouflon: (SW Asia) SW U.S.
O. dalli, Dall's Sheep: Alas., NW Can.
O. canadensis, Bighorn Sheep: SW Can. to N Mex.

About the Authors

Ronald M. Nowak (INTRODUC-TION and ALIENS) received his Ph.D. in biology from the University of Kansas in 1973. He was staff mammalogist at the U.S. Interior Department's Office of Endangered Species from 1974 to 1987. He is author of the forthcoming 6th edition, as well as two previous editions, of *Walker's Mammals of the World* and of many other publications on mammals and wildlife conservation.

Don E. Wilson (MARSUPIALS, INSECTIVORES, and XENARTHRANS) made his first trip to the tropics in 1964, as an undergraduate; he has returned yearly to conduct research on mammals and to teach tropical biology courses. Formerly editor of the *Journal of Mammalogy*, he also serves on the editorial boards of several conservation and biological organizations, and has written over 150 scientific publications, including books on mammals of the world, mammals of New Mexico, and bats. Currently, he is the director of biodiversity programs at the Smithsonian Institution's National Museum of Natural History.

M. Brock Fenton (BATS) is a professor of biology at York University in Toronto. Most of his research has involved bats, which he has studied in the field in North and Central America as well as in Australia, Africa and the Middle East. He has written four books on these fascinating animals.

Andrew T. Smith (LAGOMORPHS) is a professor of biology at Arizona State University, focusing on conservation biology, behavioral ecology, population biology, and mammalogy. Since 1969, his studies of pikas have taken him to the Rocky Mountains and the Sierra Nevada, as well as the Tibetan Plateau. He has chaired the Lagomorph specialist Group of IUCN's Species Survival Commission since 1991.

Numi C. Mitchell (RODENTS) is a biologist whose research normally focuses on the ecology of rare and endangered terrestrial vertebrates. By following animals in the wild, she identifies the critical resource requirements of rare species and uses this information to develop management strategies that protect the animals and their most desirable habitats. She currently works at the Conservation Agency, in Jamestown, Rhode Island.

Kenneth S. Norris (CETACEANS), a professor emeritus of natural history at the University of California, Santa Cruz, is one of America's senior marine mammalogists. He has spent 40 years pioneering studies of cetaceans and teaching graduate students. He also helped found the Society for Marine Mammalogy, a major international group, and served as its first president.

John Seidensticker (CARNIVORES) is curator of mammals at the National Zoological Park, Smithsonian Institution, affiliate professor of biology at George Mason University, and chairman of the Save the Tiger Fund Council. He has conducted radio-tracking studies on species in each of the North American carnivore families. He is the author or editor of 130 articles and books.

Susan Lumpkin (CARNIVORES) is director of communications for Friends of the National Zoo and editor of *ZooGoer*. She is an authority on reproductive behavior in mammals and birds, and on the maintenance of mammals in zoos. She has written or edited several books on mammals.

Burney LeBoeuf (PINNIPEDS) is professor of biology at the University of California, Santa Cruz and an authority on marine mammals. He is best known for his studies of elephant seals dealing with their social and reproductive behavior, as well as their diving and foraging behavior.

Galen Rathbun (MANATEES) is a wildlife research biologist with the Western Ecological Research Center of the U.S. Geological Survey. In addition to studying manatees, he has worked with elephant-shrews in East Africa, dugongs in Australia and Palau, and sea otters in California. He continues to research declining species and their habitats, focusing on terrestrial animals of California, including San Joaquin antelope squirrels, giant kangaroo rats, California red-legged frogs, and western pond turtles.

Valerius Geist (UNGULATES) is a professor emeritus of environmental science at the University of Calgary, in Alberta, Canada. Best known for his early work on North American mountain sheep, he has spent nearly four decades studying the behavior, ecology, and evolution of ruminants, with emphasis on Ice Age speciation. He is the author of 13 books, including some on human evolution during the Ice Age. Most of his scientific work was done in wilderness areas of northern Canada.

Acknowledgments

The Book Division is indebted to many wildlife researchers and others for their assistance. Special thanks goes to Michael D. Carleton, curator, Division of Mammals, at the Smithsonian Institution's National Museum of Natural History, and to Lyn Clement for her editorial services.

Additional Reading

The reader may wish to consult the *National Geographic Index* for related articles and books. In addition, the following sources may be of interest: *Walker's Mammals of the World*, 5th edition, by Ronald M. Nowak; *The Mammals of North America* by E. Raymond Hall and Keith R. Kelson; *The Mammals of Canada* by A.W.F. Ganfield; *Mammalogy* by Harvey L. Gunderson; *National Audubon Society Field Guide to*

North American Mammals by
John O. Whitaker, Jr.; *Mammals
of the Carolinas, Virginia, and
Maryland* by Wm. David Webster,
James F. Parnell, and Walter C
Biggs, Jr.; *Bats* by M. Brock Fenton;
Introducing the Manatee by
Warren Zeiller; *The Natural
History of Seals* by W. Nigel
Bonner; *The Natural History of
Whales & Dolphins* by Peter G.H.
Evans; *Sierra Club Handbook of
Seals and Sirenians* by Randall
Reeves, Stephen Leatherwood,
and Brent Stewart.

Picture Credits

Cover, Michio Hoshino/Minden
Pictures. 1, Art Wolfe. 2-3, John
Garrett/Tony Stone Images. 4-5, Jeff
Foott/Bruce Coleman, Inc. 6-7, Daniel
J. Cox/Natural Exposures. 11, Ken
Highfill, The National Audubon
Society Collection/Photo Researchers,
Inc. 12-13, Steve Gettle/ENP Images.
16, Art Wolfe. 17, Steven Holt/Aigrette
Photography. 18-19, Rod Planck, The
National Audubon Society Collection/
Photo Researchers, Inc. 22, Wayne
Lankinen/ Bruce Coleman, Inc. 23,
Zefa/The Stock Market. 24, Dwight R.
Kuhn/Bruce Coleman, Inc. 25, John R.
MacGregor/Peter Arnold, Inc. 26-27,
Merlin D. Tuttle/Bat Conservation
International. 32 (upper left), Wendy
Shattil/Bob Rozinski; (all others),
Merlin D. Tuttle/Bat Conservation
International 33 (upper left), Gilbert
Grant, The National Audubon Society
Collection/Photo Researchers, Inc.; (all
others), Merlin D. Tuttle/Bat Conser-
vation International 35-37 (all), Merlin
D. Tuttle/ Bat Conservation Interna-
tional 38, Bianca Lavies/NGS Image
Collection. 41, Bianca Lavies. 42-43,
Tom & Pat Leeson. 48-49, Michael S.
Quinton. 50, Art Wolfe/Tony Stone
Images. 51, Irene Hinke-Sacilotto.
52-53, Ron Sanford. 59, David Welling.
60 (left), Daniel J. Cox/Natural Expo-
sures; (right), Robert E. Barber. 61 (left),
Laurie Campbell/Tony Stone Images;
(right), John Cancalosi. 62, Jeff Foott/
Bruce Coleman, Inc. 63, Bob & Clara
Calhoun/Bruce Coleman, Inc.
64, Daniel J. Cox/Natural Exposures.
64-65, Tom & Pat Leeson. 66, W. Perry
Conway. 67, Wendy Shattil/Bob
Rozinski. 68-69, Tom & Pat Leeson.
70-71, Michael Melford/The Image Bank.
76-77, Flip Nicklin/Minden Pictures.
78-79, Erwin & Peggy Bauer/Bruce
Coleman, Inc. 79-81 (both) Flip
Nicklin/Minden Pictures. 82, Art Wolfe.
83, Flip Nicklin/Minden Pictures.
84-85, Brandon D. Cole/ENP Images.

85, Daniel J. Cox/Natural Exposures.
86-87, Ron Sanford/Tony Stone Images.
88-89, Flip Nicklin/Minden Pictures.
89, Daniel J. Cox/Natural Exposures.
90-91, Daniel Larsen/Bruce Coleman,
Inc. 92-93, Art Wolfe/Tony Stone
Images. 99, Wendy Shattil/Bob Rozinski.
100, Michael S. Quinton/NGS Image
Collection. 100-101, Wendy Shattil/Bob
Rozinski. 102-103, Walt Enders/ENP
Images. 103, Joe Van Os/The Image
Bank. 105, Tim Davis/Davis-Lynn
Images. 106, Michio Hoshino/Minden
Pictures. 107, Erwin & Peggy Bauer.
108, Jim Stamates/Tony Stone Images.
109, Gary Braasch/Tony Stone Images.
110-111, Daniel J. Cox/Natural Expo-
sures. 112-113, John Shaw/Bruce Coleman,
Inc. 112-113, David Madison/Bruce
Coleman, Inc. 114-115, Ted Levin. 116-
117, Michael S. Quinton/NGS Image
Collection. 118, Daniel J. Cox/Natural
Exposures. 119, Daniel J. Cox/Tony
Stone Images. 120, Alan & Sandy
Carey. 121, Tom Walker/Tony Stone
Images. 123, Tim Davis/Davis-Lynn
Images. 124, W. Perry Conway. 124-
125, Robert E. Barber. 126-127, Jean
Stoick/Peter Arnold, Inc. 128, Daniel J.
Cox/Natural Exposures. 129, Renee
Lynn/Davis-Lynn Images. 130-131,
Thomas Kitchin/Tom Stack & Associ-
ates. 137, David Woods/The Stock
Market. 138-139, Tom & Pat Leeson.
140-141, Tim Davis/Davis-Lynn
Images. 141-143 (both), W. Perry Con-
way. 144-145 (both), Frank S. Balthis.
146-147, Douglas Faulkner/The Stock
Market. 150, James D. Watt/Pacific
Stock. 151, Douglas Faulkner/The Stock
Market. 152-153, Jeff Lepore, The
National Audubon Society Collection/
Photo Researchers, Inc. 158, Larry Ditto/
Bruce Coleman, Inc. 159, Alan & Sandy
Carey. 161, Daniel J. Cox/Natural
Exposures. 162-163, Tom & Pat Leeson.
164-165, Michio Hoshino/Minden
Pictures. 166, Robert E. Barber. 166-
167, Art Wolfe. 168-169, Tony Dawson/
Tony Stone Images. 169, Kennan Ward/
The Stock Market. 170-171, Michael
Durham/ENP Images. 173, William
Albert Allard/NGS Image Collection.
174-175, Ted Levin. 175, Tom & Pat
Leeson. 176-177 (both), Paul McCormick/
The Image Bank. 178-179, Laguna
Photo/Liaison International. 180,
Wendy Shattil/Bob Rozinski. 180-181,
Tom Boyden. 182-183, Jeff Lepore,
The National Audubon Society
Collection/Photo Researchers, Inc.
184-185, Ralph Perry/Tony Stone
Images. 188-189, John Eastcott & Yva
Momatiuk/NGS Image Collection.
189, Jane Burton/Bruce Coleman, Inc.
190, Joe & Carol McDonald/Bruce
Coleman, Inc. 191, Renee Lynn/Davis-
Lynn Images.

Library of Congress ℗ Data

Wild animals of North America / prepared by the Book Division. National Geographic Society.
     p.   cm.
     Includes index.
     ISBN 0-7922-7062-2 (reg). —ISBN 0-7922-7066-5 (dlx)
     1. Zoology—North America.
     I. National Geographic Society (U.S.) Book Division.
     QL151.W55 1998
     591.97—dc21   98-2798
                           ℗

Composition for this book by the National Geographic Society Book Division. Color separations by Quad Graphics, Martinsburg, West Virginia. Printed and bound by R. R. Donnelley & Sons, Willard, Ohio. Dust jacket printed by Miken, Inc., Cheektowaga, New York.

Visit the Society's Web site at www.nationalgeographic.com.